The Science of Sympathy

HISTORY OF EMOTIONS

Editors
Susan J. Matt
Peter N. Stearns

A list of books in this series appears at the end of this book.

supported through grant
Figure Foundation
on the origin of emotion

The Science of Sympathy

Morality, Evolution, and Victorian Civilization

ROB BODDICE

University of Illinois Press
URBANA, CHICAGO, AND SPRINGFIELD

1 2 3 4 5 C P 5 4 3 2 1
♾ This book is printed on acid-free paper.

The cataloging-in-publication data is available
on the Library of Congress website.

ISBN 978-0-252-04058-0 (Hardcover)
ISBN 978-0-252-08205-4 (Paperback)
ISBN 978-0-252-09902-1 (e-book)

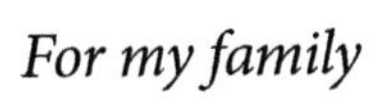
For my family

Contents

Contents

Acknowledgments

This scope of this book has grown enormously since its beginnings as an inquiry into the motivations of mid-Victorian physiologists. Up until 2011 I had been working on the history of cruelty and callousness, and on various expressions of the "man of feeling," but I had not yet discovered (perhaps because it did not yet substantially exist) the "history of emotions" as a theoretical and methodological organizing principle. It is safe to say that this new field of inquiry in the discipline of history has since taken off, expanding at a truly remarkable rate, and I have been fortunate to have been among an international network of historians pushing the boundaries of the possibilities of research into emotions of the past. At some point or another, research in progress related to this book has been tested at the European Society for the History of Science conference in Athens; the Montreal British History Society; the North American Conference on British Studies; the "Feeling Space" workshop at the University of Copenhagen; the *Centro de Ciencias Humanas y Sociales* in Madrid; the Department of History at the Universität des Saarlandes; the Society for the Social History of Medicine conference; the Institute of Historical Research London; the Centre for the History of Emotions at Queen Mary, University of London; the Department of History, Classics, and Archaeology at Birkbeck, University of London; and at Bath Spa University under the auspices of a Royal Historical Society conference on "Intimacy, Authority and Power." My thanks respectively to Katalin Straner, Brian Lewis (twice), Karen Vallgårda, Javier Moscoso, Wolfgang Behringer, Rhodri Hayward, Kate Bradley, Thomas Dixon, Lesley McFadyen, and Bronach Kane. The research and writing has principally taken place at the Cluster of Excellence Languages of Emotion at Freie Universität Berlin and at the Center for the History of Emotions at the Max Planck Institute for Human

Development in Berlin. The work owes a large debt to these institutions, and to their principal heads: Hermann Kappelhoff and Ute Frevert.

Archival research was carried out in London at the Wellcome Library, the British Library, University College, the Linnean Society, and Imperial College; in Berlin at the Humboldt Grimm-Zentrum; and in Boston at the Countway Library for the History of Medicine. My thanks to all. Some of the material in this book has been extracted and repurposed from the following articles: "German Methods, English Morals: Physiological Networks and the Question of Callousness, c.1870–81," in *Anglo-German Scholarly Relations in the Long Nineteenth Century*, ed. Heather Ellis and Ulrike Kirchberger (Leiden and Boston: Brill, 2014); "Species of Compassion: Aesthetics, Anaesthetics and Pain in the Physiological Laboratory," *19: Interdisciplinary Studies in the Long Nineteenth Century* 15 (2012). I thank the Wellcome Library, London, for their generous policy regarding the use of images.

The emergence of centers of emotions history research notwithstanding, much of the book's theoretical and methodological substance comes from foundational works in this discipline that have endured. In particular the work of William Reddy has made this book possible, and I thank him also for his personal support. The book has been finished under the auspices of a research grant from the Deutsche Forschungsgemeinschaft, based at the Friedrich-Meinecke-Institut at Freie Universität Berlin. My thanks especially to Martin Lücke and to Eva-Maria Silies for being instrumental in making this happen. I owe a debt of gratitude to Jan Plamper that I doubt I shall ever be able to repay. Thanks to the series editors, Susan Matt and Peter Stearns, and to Laurie Matheson at the University of Illinois Press, for seeing this work through to publication, and to the two anonymous reviewers for their positive endorsement of the manuscript and for their insightful criticisms.

Further acknowledgments are due to some wonderful scholars who have selflessly given their time, their thoughts, and their friendship to me in the processes of research and writing. Important interventions by Daniel Brückenhaus and Philipp Nielsen have saved me from German calamity. Joanna Bourke is a constant source of inspiration. Brian Lewis and Elizabeth Elbourne have become my Montreal pillars of wisdom. Michèle Cohen continually gives up her front room for the sake of my London research, and regales me with tea and not a little sympathy. Finally to friends—too many to mention—and family, my deepest thanks for continuing to assure me that this is all worthwhile. To Tony Morris for embracing my work and my career with a truly welcome dose of enthusiasm and professionalism. And to my wife, Stephanie Olsen, for being my second brain, tirelessly interested, and forever saving me from the worst of my literary and intellectual mistakes.

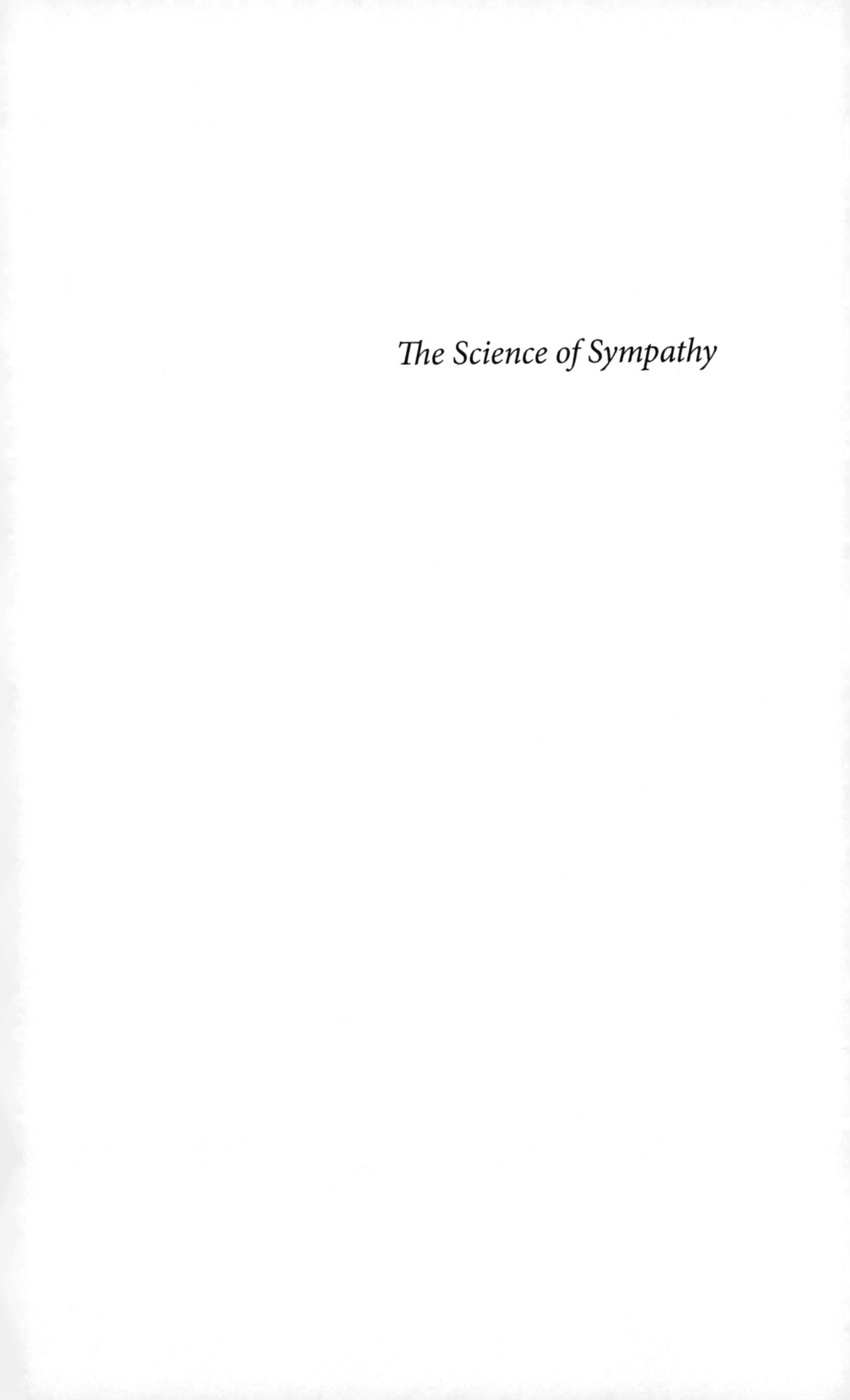

The Science of Sympathy

1 Emotions, Morals, Practices

Domestication of the Human

Everyone has some sense of the outrage and controversy wrought by the theories of Charles Darwin. Whether from the perspective of religious convictions, scientific theories, or moral codes, the scope of the intellectual impact of Darwin's theories is well known. But what did people actually *do* with those theories? I ask this in the most practical sense. How were Darwin's ideas translated into scientific, medical, and social practices?

This book is about only one of those ideas, but a central one. In Darwin's *Descent of Man,* published in 1871, he asserted that the basis of morality in civilized societies was derived from a naturally occurring and highly developed capacity for sympathy.[1] The more of it a population had, the more moral and the more civilized it would be. It was the central pillar in an argument constructed to justify the sense of superiority commonly felt by the inhabitants of Darwin's own world. Through a highly civilized sympathy, a compound of moral sense and intellect, Darwin explained the superiority of white-skinned, upper-middle-class, educated Victorian men.[2] As "love, sympathy and self-command" were "strengthened by habit," aided by superior reasoning, Darwin's sympathetic man would declare himself the supreme judge of his own conduct, swearing an oath against the violation of the "dignity of humanity." No "barbarian or uncultivated man could thus think."[3]

To demonstrate this relationship between sympathy and morality, Darwin had to show that thinking and feeling—both qualities determined by nature—led to acting in certain ways. He provided an intellectual blueprint for why Victorians thought, felt, and commonly acted in virtuous ways, according to the standards of virtue of the time. To that extent, Darwin himself demonstrated how ideas led to practices. Numerous examples were given, which I will explore in some detail below. A brief case will introduce the principle. A man jumps into a river to save a drowning boy, even though the boy is unknown to him. Or perhaps it is not a boy, but a dog. Darwin suggested that the extension of feelings of sympathy beyond the immediate kin group, or even beyond the species barrier, enabled civilization to flourish. Things done in the name of this sympathy—the saving of a boy's life, or a dog's life—were good for the community writ large. The action, which we might call altruistic, a word not used by Darwin, did not need to be intentionally carried out.[4] It followed instinctively from the rising feeling of distress at the suffering of another. This instinctive sympathy, combined with an instinctive altruistic act, was a building block for civilization. The more people in an extended community who felt and acted in this way, the more likely that community would survive. It was a social vision of "the survival of the fittest."[5] Individual lives were bound up in a web of sympathy and the actions that corresponded to it. Over time, those actions were intellectualized: An empirical understanding of their goodness led to them being labeled as moral. Thereafter, such morals are prescribed and taught, but it was essential for Darwin that the attempts of "public opinion" to nurture such morals *followed* the existence and evolution of sympathy according to natural law.

Darwin's vision was natural-historical. He was looking back to explain the present, and had only half an eye on the future. Yet the admission that latent qualities could be nurtured, subjected to abstract reason and learned judgment, seemed also to imply a way forward. Darwin's work on evolution had, since the publication of *Origin of Species* in 1859, used domestication as a metaphor for natural forces. Breeding for specific qualities in fancy pigeons, for example, the expert breeder could change the form of the bird quite rapidly, doing in a handful of years what "nature" would do over thousands.[6] The suggestion implicit in *Descent of Man* was that human beings would respond similarly to such programmatic "domestication." Victorian watchwords such as "progress" and "improvement" could be planned at the societal level by modifying the human. Those at the apogee of evolutionary forces—men like Darwin himself—could harness their powers of reason to enhance their natural sympathetic qualities. They could also use the instruments of public opinion, using artifice to improve the sympathetic movements of society in general. With the dawn of a new scientific revolution, based on evolutionary theory and experimental endeavor, a new blueprint for sympathy could be envisaged, and with it, a new definition of morality for the actions carried out in the name of sympathy.

Specifically, then, this book is about the translation of Darwin's ideas on the evolution of sympathy and its relationship with morality to the everyday professional and affective practices of the first generation of Darwinists. I begin with the premise that sympathy is not a stable thing.

What is Sympathy?

If I were to ask "what is sympathy?" now, I would get a range of confused and confusing answers. The word *sympathy* and its derivatives are employed in a dizzying range of rhetorical contexts that obfuscate its meaning. The word derives from the Greek and means, literally, to suffer *with*. In this sense, it is exactly synonymous with its Latin translation, which has come down to us as "compassion." The Greek *pathos*, much like the Latin *passio*, both indicate suffering, but in a neutral sense. One suffered love as much as one suffered pain. The two words have diverged a little, compassion now being commonly associated with suffering in the modern sense, whereas one might still use the word "'sympathy" to express a shared positive emotion, such as joy or love. Still, there is much semantic confusion, which adequately represents the conceptual ambiguity of sympathy. An etymological study will find sympathy and compassion to be synonymous with "pity," "tenderness," and "humanity." It is essential to begin with this conceptual mess, for conceptual messes make history. Sympathy provides many bases on which to build arguments and forge practices. From the point of view of those on different bases, the arguments and practices of others are bound to seem wrong, but all accusations of wrongness would seem to the accused to have come from false premises.

When acknowledging the grief of a friend, we may send a card that reads "In sympathy." An angry person complaining of injustice may rouse us to "sympathize with her point of view." We read a novel and have an implicit understanding of whether a character is sympathetic or unsympathetic. If you break your leg I may sympathize with you, or I may offer sympathy, which may or may not be the same thing. If I broke my leg I might hope for sympathy from others, because it might make me feel better. I do not need to theorize how this palliative works, I simply know that it does. Sympathy is existential (I *am* sympathetic), emotional (I *feel* sympathetic), practical (I *act* sympathetically), and affective (I *seem* sympathetic). In a given instance it has surface and/or depth and/or consequence. If we highlight the emotional and affective elements here, things get more complicated still. I am sympathetic *with* your anger, or at least, I *seem* to be. Does this feel the same or seem the same if I am sympathetic with grief, joy, suffering, pain, depression, or hope? Sympathy then seems to be a vehicle for translating the emotions of others, rather than an emotion in itself. It is an enzyme, if you will, a catalyst that converts external experience into internal experience, and helps to fashion responses. Yet it seems also to be an emotional disposition, as well as a discrete emotion itself. It is the tingling feeling you get when hearing of another's painful injury, and your capacity immediately to access whatever the emotional context is. In short, in our own diction (the community here is Anglophone—it would be even more complex if the linguistic net were broadened) sympathy means many things, some apparently contradictory. We know, implicitly, what we mean when we use the word, but it eludes us when we try to define it.

This elusiveness is not new. An awareness of it, or at least a moment's reflection on the difficulty of describing this state of being, this emotional category, this emotional mechanism, will help to make sense of this book's inquiry. It would be a commonplace for me to say that, despite the difficulty with which we might attempt to describe sympathy in the here and now, we would nevertheless readily admit that we carry out certain actions *because of* or *through* sympathy. Indeed, many of the things we *do* as an effect of sympathy would be held up as ways of defining what sympathy is as the cause. I give money to a certain charity as an act of sympathy for whatever or whoever is the object of that charity. Sympathy is therefore an emotional recognition of the plight of the object in question. I give a friend a hug after the funeral of his brother. The act demonstrates the feeling of sympathy. I hide my face behind a cushion when watching an embarrassing moment on television. Even though I do not share a real space with the origin of the embarrassment, my actions seem to describe a sympathetic activity in me. The argument that I will develop here is quite simple. The things we do, the actions we make that indicate sympathy, change over time. They are changed both consciously and unconsciously according to shifting notions of what sympathy is and, connectedly, what *ought* to be done because of it.

The introduction of the word "ought" signals the fundamental quality of sympathy that I have yet to mention. It is moral. The actions done in the name of sympathy are, or are supposed to be, good. From Hume and Smith down to Jesse Prinz, the idea that sympathy is the basis of morality has been a central pillar of ethics.[7] The details of this relationship have proven to be difficult to agree upon, but the essentials of it have not. But a clue as to the historical variability of sympathy and its meanings immediately becomes clear. The history of morality is well developed in various models, from Norbert Elias's "civilizing process" to M.J.D Roberts's *Making English Morals*; from histories of Puritanism to evangelicalism to humanism.[8] That morals have changed over time—to be clear, that right and wrong have had different meanings in different times and different places—is widely accepted as an obvious observation. We see it in the eradication of torture and capital punishment from the justice system; the removal of corporal punishment from the education system; the waning influence of established churches over the rectitude of, for example, sex before marriage, promiscuity, and homosexuality; and the British abandonment of the slave trade. What was once unthinkingly accepted as right is now considered wrong. In our own times, liberal thinking constantly pushes us to re-evaluate our commonplace assumptions and unthinking attitudes, particularly as concerns the implicit structures of sexism, racism, classism, and speciesism. Our morals are in a state of flux. This is not, or at least need not be, a narrative of progress. Earlier historians of morality might have made a Whiggish stand, but it is not necessary. Moral standards change according to the people and institutions who have the power to change them. We judge our contemporaries by our own moral standards. Any historian will tell you that it would be foolish to judge historical actors by anything other than *their* moral standards. The notion of moral change is implicit to historical work.

If that is true, and I expect little dissension on that point, then it must follow that being and feeling sympathetic and acting sympathetically must have changed too. For some reason, this has proven to be a much more difficult sell than the history of morality. Until recently, historical work on the emotions has been overshadowed by those disciplines that might be considered the "natural" home of emotions research. Psychology, anthropology, and neuroscience have spun their own threads on the science of emotions, with techniques and language that seem at first to disqualify the historian from engaging. This is not our field of expertise. We are dabbling laymen. We do not have the required sophistication. These feelings might dwell among historians themselves, wary of entering unto "soft" subjects that defy empiricism.

This is beginning to change, but the early moves in the history of emotions have inevitably seemed like iconoclastic violence. Since change-over-time is the *metier* of the historian, and the existence of a set of basic and universal emotions is the principal claim of some emotion scientists, historians have immediately set about toppling this transhistorical statue, shattering it into relativistic pieces. Inevitably there has been antagonism. More often, historians have simply been ignored. If we are seriously to demonstrate that emotions have a history, the challenge of empiricism has to be met. How does one demonstrate that emotions in the past felt different?

My proposal, in this instance, is to do with the past what we do in the present. We measure feelings by actions, by the practices they produce. We label a moral action sympathetic and thereby define, by tracing backward, the sympathetic impulse. The past readily affords us access to actions that were declared moral or immoral. Sometimes, and this is key, whether or not an action was moral was a matter of debate. Sometimes the debate about morality was framed precisely by a discussion of sympathy. When we find actions being defined as moral because they were activated by sympathy, we can find a historical definition for sympathy. Moreover, we can overlay this backward trace with contemporary accounts of what sympathy is and where it comes from. As will become clear, the sympathy I will describe here, with its accompanying morality, is completely distinct from anything we might recognize as sympathy today. Even with the confused and confusing array of definitions and semantic difficulties, it is clear that a historical investigation into the emotional world of the past—the historical nexus of sympathy and morality—can disrupt and defamiliarize concepts, emotions, and moral categories that we perhaps take for granted. I am going to argue that, in a certain place and a certain time, vivisection was an act of sympathy. I am going to argue that legally compelling parents to vaccinate their children on pain of imprisonment with hard labor was an act of sympathy. I am going to argue that the blueprints for a society built on a eugenic creed was borne of a certain understanding, being, and feeling of sympathy. And I am going to argue that all these things, in that time and place, and because of the unfamiliar, now lost, sympathy that activated them, were *moral*.

Recipe Knowledge: New Practices, New Morals

The subject of this book is a loose community of "scientists," incorporating evolutionary theorists, physiologists, toxicologists, medical doctors, public health officials, political figures, biometrists, vaccinists, and eugenicists. What they have in common is, at the very least, an intellectual heritage that begins with Charles Darwin's *Descent of Man*, published in 1871. Many of them have much more than this in common with Darwin, from direct correspondence to friendship to loyal advocacy of his theories. They represent the first generation of Darwinists, before the onset of Neo-Darwinism, and before Darwin's "ism" was really an established phenomenon. Darwin was one among many evolutionary theorists, though admittedly the most important, but the observation serves as a reminder that we should not expect to find a uniformity of thinking among early Darwinists. They invoked Darwin's name and referred to his theories insofar as they served their own ends, making for a messy and pluralistic array of evolutionary principles and practices. Importantly though, all of these figures utilized, explicitly, a theory of sympathy constructed by

Darwin and worked out how to apply it, live it, and embody it. Their interpretations of Darwin's sympathy were also pluralistic. Evolutionary scientists were as wont to talk past each other then as they are today. Nevertheless, from about 1870 until the early years of the twentieth century, an inchoate discourse of sympathy sprang up, displacing, or attempting to displace, the prevailing understanding of sympathy. The moral universe of Adam Smith and David Hume, not to mention the charitable sentiments of the Lady Bountiful, were considered unfit for a world understood by evolutionary principles and natural law. This was not just an intellectual posture, but an attempt to alter the feelings and judgments of influential men (mostly men, at any rate). The alteration was necessary. If scientific and medical progress was to be built upon animal experimentation, for example, and if progress was incompatible with cruelty or callousness, then animal experimentation could not be allowed to fit into common ascriptions of cruelty and callousness. Vivisection, the good outcomes of which were shouted from on high, must have had moral means if it had moral ends. Cutting the frog, the rabbit, and the dog had to be acts of sympathy because the desired outcome was a diminution of suffering. Similarly, if vaccination put an end to smallpox, but parents refused to vaccinate their children, then it was moral to punish them. Their actions risked the suffering or death of the community. Compulsion was an act of sympathy. And if the good of society depended on "fitness," then was it not an act of sympathy to discourage or prevent the "unfit" from breeding, and to encourage the "fit" to procreate? These lines of argument were not universally followed, or even commonly assented to. But among communities of involved scientists they became accepted truths, to the point that practice and theory matched each other. A Darwinian idea about the origins of sympathy had given rise to a set of Darwinian practices of sympathy, which ran in a dynamic relationship with Darwinian feelings of sympathy. I have limited the focus to these three particular case studies because they offer the most direct evidence of the conscious employment of evolutionary thinking in new medical and scientific practices. Of course, a great many other examples come to mind (some of which are hinted at in chapter 2), both within the scientific world, from new innovations in surgery to sanitation reform, and beyond it, from the policing of the Empire to the reform of education policy.

My point is not to shock, nor to condone, but simply to explain. These things, vivisection, state compulsion, and eugenics, have, at best, an air of ethical dubiousness about them in contemporary Western society. Some would argue forcefully that there is a measure of evil in all three. How can I set out to describe them as moral? From where did those practices derive their moral purchase? How did it feel to carry out these practices?

It is not so difficult. First of all, I am not a moral relativist from a philosophical point of view. I make moral judgments in my own life as readily as the next person. But I am prepared to acknowledge that when I do make moral judgments, or when I am myself held up to moral standards, that these are not absolute. They work in a given setting that brings order, sense, and meaning to my life and the lives of others around me. Morals work in context and are, for want of a better word, functional. I am, therefore, a moral relativist from an historical (or cultural) point of view. This is only to acknowledge that morals are constructed—they function—differently in different times and places. Anybody who has ever planned a trip to a foreign country knows the essential truth of this. It is so manifestly obvious that it does not require further justification. The past, however, notwithstanding its description as a "foreign country," is more difficult to visit. Moreover we might be reluctant to acknowledge that times relatively close to our own could have harbored such different ways of thinking, feeling, and acting.

The answer to this problem is old-fashioned historical research. I am not setting out to claim that *everyone* thought vivisection, compulsory vaccination, and eugenics were moral. I am setting out to show how certain communities, who happened to have a substantial public presence, could be convinced of this. Indeed, I aim to show that they remained convinced even in the face of fierce opposition. I aim to show how this way of thinking, feeling, and acting came about for this community. This completes the picture sketched above. As well as working back from actions called moral in order to find a definition of sympathy, I also work forward from an intellectual blueprint for what sympathy is and locate it in the actions done in its name. This is nothing but traditional empirical research. The intellectual history of sympathy is recorded, like anything else, in the historical record. What people did in its name is likewise recorded. The two, at least for the period I am considering here, have never been joined up.

In a path-finding article of 1985 in the *American Historical Review*, Thomas Haskell set out to determine why the "humanitarian sensibility" emerged in the eighteenth century, particularly with regard to the campaign to end the slave trade, but also to try to understand why that humanitarian sensibility was unable to see or act upon the continuing inhumanity of wage slavery much closer to home.[9] The problem he presents is not so much historical as ethical: Why did the abolitionists, and why do we, happily ignore most of the misery in the world, even while we acknowledge that misery exists? "What enables us all [. . .] to maintain a good conscience, in spite of doing nothing concretely about most of the world's suffering, is not self-deception but the ethical shelter afforded to us by our society's conventions of moral responsibility."[10] This "ethical shelter" is going to look different in different times and places, and will change its appearance as conventions change. Adam Smith had worked this out in his *Theory of Moral Sentiments*, noting that for sympathy to function in an individual, the justness of the suffering of another had to accord with the potential sympathizer's own feelings. Where the witness to suffering finds the sufferer's lamentations to be disproportionate, or their cause mysterious, sympathy is withheld. Likewise, when the suffering is distant, abstract, and happening to people we do not know, we may lack the conventions to bring home this suffering to ourselves. Thus, though we acknowledge that suffering exists, we are not moved to do anything about it.

Haskell provides a model for historical change and the extension of sympathy or humanitarian sentiment to objects formerly considered to be outside the range of ethical concern, or at least beyond the range of ethical action. This change is not brought about simply by "the proliferation of sermons and other texts on the importance of love and benevolence," because such a proliferation only begs the question in a new way. Rather, what is required is "an expansion of the range of opportunities available to us for shaping the future and intervening in other lives."[11] Referring to "recipe knowledge," that is, the capacity to know how to make something happen from a given set of "ingredients," Haskell suggests that humanitarian action can only take place where actors perceive that certain causal chains are possible. If it cannot be perceived that certain acts can bring about certain events, then there can be no feeling of moral responsibility. On the other hand, new recipes for making things happen "can extend moral responsibility beyond its former limits."[12] "As long as we truly perceive an evil as inaccessible to manipulation—as an unavoidable or 'necessary' evil—our feelings of sympathy, no matter how great, will not produce the sense of operative responsibility that leads to action aimed at avoiding or alleviating the evil in question."[13] Evils become accessible to us only when "recipes for intervention" are of "sufficient ordinariness, familiarity, certainty of effect, and ease of operation that our failure to use them would constitute a suspension of routine, an out-of-the-ordinary event, possibly even an intentional act in itself."[14]

This approach, aimed at understanding the rise in public discourse of sympathy in the eighteenth century, with its tangible effects on the slave trade, is also of central importance to the argument laid out in this book. While the "age of sensibility" in the eighteenth century, which encompasses sympathy, has come to be well understood, it is the contention of this book that the meaning and practical implications of sympathy underwent a wholesale change from the mid-nineteenth century onward.[15] Whereas the eighteenth-century rise in humanitarian sensibility has been deployed to make sense of the antislavery movement, the animal welfare movement, drives to improve public morals, and changes in social and reading behavior, the notion of a similar emotion- and morality-driven causality in the nineteenth century has been largely overlooked. The era of scientific experimentation, of the birth of social medicine, and of the inauguration of a eugenics creed can all be explained by new "recipe knowledge," based on an idea of practical sympathy that had not been available to past actors. Moreover, this new recipe knowledge was only available to a select few who had access to both specialist knowledge and specialist practices effectively to alleviate pain and suffering, in their own terms, with a sense of "ordinary" procedure.[16] To those who did not have access to this specialist knowledge, those who inherited Smith's sympathy, this new sympathy and the new morality of evolutionary science inevitably would have seemed like an abomination.

To stick with Haskell's notion of an "ethical shelter," affording us comfort while we do nothing, it is possible to explain the emergence of differences of opinion on moral matters. Where societal conventions of moral responsibility change or are destabilized through access to new recipe knowledge, the ethical shelter upon which we formerly relied now starts to leak from the roof. A new shelter must be built, which fits the new conventions, or else the foundations of the old shelter need to be shored up so that the roof may be safely fixed. Where a society's conventions seem to bifurcate, there will inevitably be conflict along moral lines. Each will view the other from his or her own respective shelter, and his or her own understanding of what must morally be done. In the eighteenth century, new moral conventions swept up literate society as a whole, since the circumstances of change—the impact of capital in a rapidly transforming economic landscape—affected everyone. I contend that the changes in the nineteenth century depended to an unprecedented degree on specialist or professional knowledge. This knowledge impacted society at large, but did not necessarily reveal its causal operations. The paradox of a scientific world that was at once public and somehow hidden in plain sight by dint of its complexity lent it an air of mystery, fear, and danger.

Nevertheless, to the experts in question, new scientific knowledge seemed to provide exactly the new kinds of recipe knowledge Haskell describes. New forms of readily accessible, ordinary behaviors led to new social goods. Physiological and toxicological research on animals led to new understandings of the function of the human body, which in turn led to medical advancements in surgery, medicine, and disease prevention. Given the potential of these new procedures to alleviate suffering, it quickly became apparent to those with this knowledge that it would be morally inexcusable not to practice them. Inaction, to use Haskell's words, came to be seen as "not merely one among many conditions necessary for the occurrence or continuation of the evil even but instead a significant contributory *cause*" of that evil.[17] To know that vaccination prevented smallpox and then consciously to not vaccinate everyone was, in effect, to permit smallpox to happen. To know that toxicological research on animals would lead to the effective treatment of diphtheria and then not do that research was effectively to leave the field for diphtheria to take its victims. To understand the statistical science of racial degeneracy and then to allow the "weak" and the "unfit" to continue to breed was in effect to commit a race crime. Conversely, to pursue these new procedures to remedy the ills of society afforded its agents an enormous sense of satisfaction. As Smith had maintained, sympathetic feelings that gave way to benevolent actions to assist the sufferer ended up satisfying the sympathizer. A broad acceptance of the assumption that the Golden Rule was always active ensured that the sympathetic would likely have sympathy returned to them in their own times of need.[18]

When physiology, germ theory, and racial statistics emerged as major institutions of modern science in the second half of the nineteenth century, they seemed to afford their practitioners ways of providing far-reaching goods combined with unprecedented degrees of scientific satisfaction. The intricacies of scientific practice, its ethical power, and its emotional impact were all rolled up together to produce the modern scientific self as a benevolent, knowledgeable, moral creature. He—and the gendering is important—came replete with a theory of sympathy that departed from the legacy of Smith, and a set of practices that could seemingly reify that theory. To his opponents, who could not understand the theory and could not access his methods, the scientist with his "science" of sympathy appeared as a dangerous moral aberration. Because those outside science did not understand the terms by which their ethical shelter was being destroyed by scientific expressions of knowledge, and by the ready-to-hand nature of new scientific methods, they fought to maintain the status quo. Experimentation on animals seemed cruel or callous; compulsory vaccination trespassed on the dignity of the family; eugenics signified a merciless attack on the poor and defenseless. The social goods purported to be at the end of these processes were not understood. The science of sympathy seemed like an abdication of sympathy. Meanwhile the scientist lambasted the self-serving sentimentalism of those who would preserve the weak, protect dumb animals while remaining blind to the diseases of humans, and administer charity willy-nilly without a thought as to just deserts. Smith had predicted such a possibility, noting that when two people could not relate to each other's understanding of suffering or how to alleviate it, they would despise each other. "We become intolerable to one another. I can neither support your company, nor you mine. You are confounded at my violence and passion, and I am enraged at your cold insensibility and want of feeling."[19]

For Smith, the question of justice was central to the understanding of sympathy. Where a person seemed aggrieved or joyful out of proportion to our sense of what they deserved to feel, our sympathy could not be engaged. We would be mystified by the excessiveness of the other's emotions and would likely only resent them. To this extent, the capacity to sympathize was connected to an individual's being embedded within social and moral conventions, and the feeling of sympathy was always subject to an appraisal—reached immediately—of the extent to which another's emotional state could be considered to be "fitting." Insofar as this involves putting ourselves in the position of the sufferer and asking ourselves what our emotions might be in the same situation, sympathy shares a process with empathy (a word of early twentieth-century coinage). The distinction is that the sympathizer does not enter into the emotions of the other. There is recognition but not identification. The actions that follow from empathy and sympathy, while possibly appearing to us equally "altruistic" (another neologism), are borne from entirely different motives.

Sympathy, for Adam Smith, is an impetus to action, and this action is moral. We witness suffering, feel sympathy, then act to alleviate that suffering. We do this, according to Smith, because we anticipate that the favor is likely to be returned under reverse circumstances. It ultimately serves a selfish motive to help another, since help obliges help in return. The Golden Rule, seen through Smith's eyes, is a mechanism by which we serve ourselves by serving others, and our capacity to serve others depends on our ability to sympathize with them. We thus enter into, take an interest in, the emotional results of other people's actions. This thesis reconciles Smith's two famous treatises, the paean to selfishness in the *Wealth of Nations* and the sympathetic imperative of the *Theory of Moral Sentiments*. Both center on self-regard as the key to success, but the *Theory of Moral Sentiments* puts that self-regard into social context to the extent that it explains how civilization works. If my best chance of success is to enter into someone else's success, then I enter into a virtuous circle or network in which my whole society succeeds. This broadly spread and mutually understood virtue of sympathetic action distinguishes civilization from the savage state. Such was the prevailing wind of sympathy, picked up by Darwin and his peers.

Moral Economies: From Intellectual History to a History of Practice

For the sake of shorthand, I refer in this book to these communities who thought, felt, and acted in certain ways, based on an understanding and practice of sympathy and morality as moral economies. The label "moral economy" has been about for a while, and has been employed for various usages. I want to reinvigorate it along the lines put forward by Lorraine Daston in an influential *Osiris* article, even though she seems to have dropped it herself.[20] In this definition, "'moral' carries its full complement of eighteenth- and nineteenth-century resonances: it refers at once to the psychological and the normative," and "economy" refers "not to money, markets, labor, production, and distribution of material resources, but rather to an organized system that displays certain regularities, regularities that are explicable but not always predictable in their details."[21] The moral economy is particularly useful for a study such as this because it encapsulates much of the jargon emerging from the history of emotions as a new field. It also resolves many of the semantic disputes that seem to be holding up empirical applications of new ideas about how to do this kind of work. I will briefly address some of these developments in order to be able to pack them away, before returning to further explore Daston's original definition of the moral economy in order to elaborate upon it. Her work predated the vast majority of thinking in the history of emotions, yet it seems to have forecast it aptly, capturing many of the concerns and resolving them with an analytical elegance that is wanting in much new work.

In particular, because I am locating certain modes of thinking, feeling, and acting within specific "communities," namely a loose community of scientists following in Darwin's wake, but spread across physiological, biological, medical, political, and public health roles, it might seem to make sense to employ Barbara Rosenwein's concept of "emotional communities."[22] I resist this mainly because of the things to which Rosenwein herself has opposed it. She devised the concept of emotional communities to make sense of the cultural conventions that define why emotional expressions seem to change from place to place and over time. Her definition was commonsensical. Emotional communities are

> precisely the same as social communities—families, neighborhoods, parliaments, guilds, monasteries, parish church membership—but the researcher looking at them seeks above all to uncover systems of feeling: what these communities (and the individuals within them) define and assess as valuable or

> harmful to them; the evaluations that they make about others' emotions; the nature of the affective bonds between people that they recognize; and the modes of emotional expression that they expect, encourage, tolerate and deplore.[23]

An individual always belongs to more than one emotional community at any given time, and moves between them. This movement, according to Rosenwein, depends on different emotional communities sharing norms, or at least being different within an acceptable degree of tolerance. There is not an emphasis here on how people felt so much as on normative codes of affective expression. This concerns the construction of "emotional standards" at various levels of society, and the ways in which these standards are employed by interrelated communities. Rosenwein is looking for the reasons for historical change in these conventions. In an attempt to cement the centrality of emotional communities in history of emotions work, Rosenwein has criticized William Reddy's concept of "emotional regimes."

Reddy's concept of the emotional regime, which was formulated prior to Rosenwein's work, essentially does the same job as the emotional community, but with a stronger emphasis on the power dynamic that gives emotional prescriptions within a given community their necessary weight. Reddy formulated the concept in relation to his major study of the end of *ancien régime* France, with the shift from a set of prescriptions for emotional expression that were heavily repressive, to a sentimental outpouring when the regime collapsed.[24] Prior to collapse, emotional expression that transgressed prescribed and culturally *enforced* norms had to take place in emotional "refuges"—usually the theater—where rules were suspended in a controlled and limited environment.

Though the analysis was convincing, Rosenwein could not disassociate the word "regime" from the modernist's focus on the nation-state, objecting that such an analytical tool was of no use for medievalists and repeating a broadside against one-eyed modernists who favor Norbert Elias's characterization of premodern times as childlike.[25] But Elias is not the central pillar of theoretical wonder for modernists that Rosenwein makes him out to be, and in any case I think she misses the enormous irony of *The Civilizing Process*. She conflates the notion of power in the word "regime" with the power inherent in a nation-state, thereby making Reddy's theory into something of limited temporal use, promoting her concept of emotional communities to the role of universal transhistorical key for unlocking the history of emotions. She claims that Reddy's tool only works where one emotional community "dominates the norms and texts of a large part of society," effectively making it a rare breed.[26] The elaboration of both concepts in this exchange, with which Reddy has not really engaged, has done neither of them a service. Concepts are useful only insofar as they are plastic to some extent. In an attempt to mark out territory within the historiography, the specific usages of both terms seems to have rendered them less useful.

It is perfectly clear that Reddy's concept of the emotional regime can apply at any level, and simply expresses the power dynamic that means a certain set of emotional prescriptions hold sway over other configurations. Reddy has tried to point this out, noting that a regime exists wherever "the sum of penalties and exclusions adds up to a coherent structure, and the issue of conformity becomes defining for the individual."[27] Rosenwein's communities do not offer a satisfactory way of understanding how power operates to influence how emotions, both felt and expressed, work. Moreover, Reddy's model offers greater scope for navigation among communities with radically different emotional norms. One can imagine radically differentiated emotional contexts between which an individual can successfully navigate, where not only structures of penalties and exclusions, but also cultural practices, define, limit or permit emotional expression in starkly contrasting ways.

The Science of Sympathy sets out to analyze just such stark emotional navigation, between worlds where vivisection was an act of humanity, to parlors where an abhorrence of cruelty to animals was *de rigueur*; from worlds built around family unity to those that advocated state intervention in private lives; from worlds of romantic love to imaginations of eugenic edicts. Such navigations were successfully accomplished, justified, and ordered, without a sense of internal contradiction or hypocrisy, though they were not without their crises. The lives of the scientists covered here were lived out in multiple regimes at once. Reddy has not given much thought to the parallel existence of different regimes, each exercising significant dominance at the same time, in the same geography, with people existing happily in both. Would not the prescriptions of one overrule the other? Would not a choice have to be made? I am not convinced that such choices were possible. Instead, compromise was sought at the same time as pressure was brought to bear to force convergence, sometimes with success, other times with failure.

This is where Daston's concept of moral economies comes to make sense, and works more efficiently than any of the neologisms devised by historians of emotions in more recent years. Daston defined the moral economy as "a web of affect-saturated values that stand and function in well-defined relationship to one another." It derives its "stability and integrity" from "'its ties to activities." The moral economy, in these terms, combines what Daston calls *Denkkollektiv* and *Gefühlskollektiv*, thought collective and emotion collective, and expresses this holistic essence through social and bodily practices.[28] The everyday actions of the scientist in the lab are dripping with affect: they are formed by and in turn are formational of affective and emotional dispositions, bound by a shared intellectual and/or theoretical rationale, and mutual reinforcement among the members. Reason and emotion are not separate entities, but bound together in a movement—*motion*—from inside to out—*emotion*—and back again. Within the history of science and the history of emotions, the history of the self is bound with the history of practice; emotional expressions and moral feelings are caught up in a reciprocal relationship with emotional cultures and moral prescriptions. We can follow Paul White's advice and find the "ways in which the emotions might be studied as objects and as agents integral to scientific *practice*: the principles of observation, experiment, and theory and, reciprocally, the practices of the self."[29] But moral economies are also framed by a dynamic of institutional power, represented in the authority of a few notable individuals, which protects them from outside attack and shores them up against internal dissent. It works as well for a laboratory as for a family; for a university or a church. It therefore encompasses the emotional community and the emotional regime, while dispensing with the messy infighting over jargon. Moreover, moral economies function like nesting dolls. The moral economy of the physiological laboratory exists within the moral economy of the scientific community at large, which in turn has to address, if not justify itself to the moral economy of society at large. Where there are tensions and strains or even apparent incommensurabilities, they have to be resolved, for internal contradiction cannot exist without causing crisis on an individual, group, or institutional level. Sometimes, the crisis occurs and a moral economy collapses: an individual breaks down or is shunned; an institution is pilloried and collapses. Other times there are modifications: society changes to meet the innovations or departures of inner, smaller dolls. Or the smaller dolls mollify the larger ones by adapting, smoothing, refining their own systems. The moral economy, or rather, moral economies, help us to understand the emotional and practical worlds of individuals in context and how they relate to other contexts in which they also take part. Moral economies also afford us the opportunity of seeing how one doll fits inside another, if you will, and what new accommodations need to be made inside the biggest doll to permit a new, small, but radical member.[30]

The structure of *The Science of Sympathy* is unusual, perhaps reflecting the unusual combination of the histories of science and medicine with the history of emotions and morality. To follow ideas into actions, the book naturally divides into distinct portions, covering the intellectual history of sympathy as Darwin and his colleagues found it in the 1870s, the cultural history of sympathy into which they injected their ideas, and finally the history of new practices (with their moral justifications), carried out in the name of the new sympathy. The order here is important. Both Darwin and Darwin's opponents started from a common reference point. Darwin's notion of sympathy can be seen as an evolutionary divergence from a thread spun by Adam Smith. Smith, in turn, had set out the basis for dominant discourses of Victorian morality, which had themselves undergone changes of a sentimentalizing and stratifying nature. I begin, therefore, by describing the intellectual origins for Darwin's account of sympathy, before discussing that account itself. This is then put into a broader context of scientific understandings of emotions in Darwin's time.

The context effectively demonstrates the fertile intellectual environment from which Darwin's seeds would germinate in a number of surprising ways. I will argue that abstract scientific reasoning about the nature of emotions led to a form of emotional work designed to embody these abstractions. For example, the notion that a higher and better form of sympathy could be reached by seeing the ultimate consequences of actions resulted in an understanding of animal experimentation as an essential procedural component of this kind of sympathy. If animal experimentation led to improved scientific knowledge or medical understanding, then society would ultimately benefit from such operations, even if the operations were painful and unpleasant (to the animal and to the scientist). The scientist, theorizing the morality of this form of practice, would invoke a spirit of sympathetic benevolence *through* practice. "As I cut, I do good" might have been the incantation. It was a form of learning to feel anew by practicing, based on an *a priori* abstract understanding of the moral worth of that practice. Ultimately, the scientist learned to redirect his feelings of sympathy away from the animal on the operating table, focusing it instead on the suffering of abstract, unknown others. The practice was already moral in theory, but came to feel moral in the process of doing it.

To see the extent to which this process departed from established norms, the second chapter deals with the world of "common compassion," by which I refer to vernacular and prevailing constructions of emotions and morals. These moral schemes were incommensurate with the new science of sympathy, unable to access it because the theory was too difficult, the practice too esoteric. Scientists, through the eyes of advocates for religious modes of sympathy and sentimental and aesthetic reactions to blood, suffering, and death, appeared to be callous in the extreme. Here I turn both to the public and vigorous debates over the morality of science and the emotional calcification of its practitioners, and to the literary embellishment of this debate in the form of the "mad scientist." The mad scientist is a figure everyone is familiar with, either through Mr. Hyde or Dr. Moreau or countless others. The better of these literary figures took their verisimilitude from their apparent closeness to real personalities. We cannot understand what was at stake in the late Victorian world, with regard to new moral practices and new emotional frameworks for those practices, without an appreciation of the nature and degree of fear that the Darwinian revolution inspired. This fear was not merely about the collapse of religion (or worse, of God), but about the collapse of the whole fabric of civilization, built as it was on a commonly understood Golden Rule inspired by sympathy.

This leads us directly into the world(s) of the scientists themselves. Having explored scientific justifications for their practices—as men of feeling and men of morals—for the public, it remains to explore how these practices were constructed and carried on, how they were justified within the scientific community itself, and the degree to which scientists disagreed about the sympathetic and/or moral qualities of certain practices. Moreover, it will be necessary to explore the locations—the spaces and places—of new scientific practices, from laboratories to administrative offices, to discover how difficult procedures could be rendered banal, mundane, quotidian, through an atmosphere of routine. My argument here will focus on the procedural repetition, particularization, and specialization of new practices in spaces built expressly for the purpose. These new "moral houses," as I will style them, gave a physical presence to the theoretical underpinning of new practices and at the same time located those practices within structures of familiarity, of routine, and of pseudo-domesticated comfort.

The importance of place was particularly acute for physiologists, whose everyday animal experiments were especially galling to their opponents. Critical to their capacity to control their sympathetic emotions in the laboratory was the employment of anesthetics. Their use was both practical and symbolic, and can stand as a metaphor for many of the processes of relearning how to feel undertaken by scientists in this period. "Aesthetics," literally, refer to physical sensibility. An aesthete *feels* the pleasure of beautiful objects. Only latterly has the word had a direct and limited association with taste, or with looking at beauty. Chapter 3 deals with this in much more detail, but suffice it to say at present that the word "anesthetic" makes much more sense in this context. It is the opposite, or the absence, of feeling. Since sympathy is literally "to suffer with," then an unfeeling *object*, such as an anesthetized animal, ought not arouse any sympathy in the observer or the operator. There could be no sympathy where there was no suffering. In itself, this was a departure, since Smith had recognized that people readily sympathized with the dead, and Spencer went on roundly to criticize those who pitied sentimental objects. Anesthetics, under this new rubric, therefore provided a chemical route to emotional control. The animal experiment was conducted for the sake of sufferers *outside the room*. There was no suffering *inside*.

The question of place was also critical in the question of compulsory smallpox vaccination. Darwin himself was a committed advocate of vaccination, which he gave as an example of the fruitful combination of medical experimentation and the evolution of sympathy for the sake of the common good. Unlike vivisection practices, which only really became common in the 1870s in England, vaccination had been carried on since the turn of the nineteenth century and had already been compulsory in England since 1853. A number of complaints dogged the increasingly radical enforcement of this law of compulsion, from the circumscription of individual liberty to the disruption of physical space that typically kept the well-to-do from physical contact with the poor. Responsibility for public vaccination was in the hands of the Poor Law Guardians, giving the vaccine the taint of the workhouse. Vaccine matter was incubated in the arms of children before being harvested and passed to further children. Fears of infection or contamination abounded, adding to the deep feeling of resentment that free choice had been denied to parents. A number of Darwin's followers entered into this debate, arguing fiercely on both sides about the practice of vaccination, backed by threat of penal law, being morally justified according to Darwinian thinking. There is a slight difference in this chapter to the previous one, for most of the scientists in question were not themselves responsible for wielding the vaccination lancet. Nevertheless, the practical grounding of Darwinian sympathy and morality in a highly public debate about an everyday practice gave vaccinists a tangible and accessible discourse to tap into. Moreover, evolutionary scientists with children had to make a moral choice about the welfare of their own children. Darwin and Huxley had their children vaccinated, with Darwin recording the details in the family Bible. Herbert Spencer, defiantly opposed to compulsion, did not have any children. Alfred Russel Wallace, the most famous of the anti-vaccinists, had his three children vaccinated, all of them before he turned to opposition. By that time, he had also had himself vaccinated, twice.[31]

The question of children—who should have them and what should they be like—is the central focus of the last chapter. Eugenics, the science of breeding good stock, has a well-known history. Despite that, its intellectual origins have been strangely overlooked. Recent work on the subject has stressed its internationalism, even at an early date, and the coming together of strands of ideas that emerged independently in different places. Nevertheless, there is a forerunner in the history of eugenics, whose influence cannot be underrated. Francis

Galton—Darwin's cousin—was the first to think through a science of human breeding: a literal attempt to domesticate Homo sapiens. There are many studies of early eugenics and Galton's influence, but most of these stories only really get going with the beginning of the twentieth century, and most of them make some argument about the influence of Darwin's thinking on Galton's theories. What is surprising, then, is that nobody has ever appreciated the importance of the Darwinian natural history of sympathy for Galton's early thought, and for the foundations of the eugenics movement. Chapter 6 sets out to recast this history to demonstrate the absolute centrality of sympathy to eugenic ideas in the first generation of their existence. This does not begin and end with Galton, but involves the development of the evolutionary understanding of sympathy from Galton to his disciple, Karl Pearson, and Pearson's own work in making a science of this sympathetic creed in the form of biometry. Moreover, Galton and Pearson's personal relationship, as well as Pearson's own family life, make for a good study of the extent to which the fathers of a eugenic creed, built on Darwinian sympathy, attempted to live according to their own standards. Even though Galton and Pearson failed in their esoteric attempt to launch a eugenic religion, they nevertheless strove to practice, personally and professionally, in accord with the theory. They strove to *be* what they imagined they could be, as scientists and as humans of the future.

Emotives and the Scientific Self

These stories of emotion work, striving, and affective practices are informed by the theoretical insights of William Reddy's concept of the "emotive." This is the theoretical underpinning of this book. Reddy famously set out an "against constructionism" position for anthropologists studying emotions, and then even more famously developed and applied this position in his book *The Navigation of Feeling*.[32] In short, Reddy rejected the social constructionism of postmodernity, arguing the senselessness of attempting to understand emotions while completely discounting the body. That said, Reddy threw no bones to the emotional universalists. The notion of "basic" or "fundamental" emotions that are transhistorical and universal among humanity do not get a look in here. Instead Reddy argued that emotions were always part of a dynamic process involving cultural and linguistic (and I would add bodily) expectations and prescriptions on the one hand, and biological "feeling states" on the other. Reddy refuses to label these feeling states with traditional emotion words, for this would miss the point. Each of us is constantly negotiating (not consciously) how we feel with how we are expected to portray feeling in a given context. This is not performance. Acting metaphors are strictly off the table. In any situation, a prescriptive code (set by the emotional regime in Reddy's view, and according to the moral economy in mine) delimits how we can express our feelings, that is, how we *emote*.

This much has been commonly understood since at least the time of Adam Smith, who knew that a person had to express his anger, joy, or fear proportionally, or according to what might be expected. Otherwise, the reaction to him would be ridicule, rejection, or hatred, instead of sympathy. Reddy basically makes the same point, but writes it large. When we emote—literally, to project outward our inner movements—we must make the outward movement fit into a predetermined model of expression. This counts for all forms of emotional expression, from language use to facial and bodily comportment, from the shedding of tears to the disciplined affective practices of quotidian life. Reddy introduces another word to cover what we emote: "utterance." Although this literally is confined to verbal expressions, I will apply it to all these forms of expression because it is much closer to what I want to suggest is taking place than words relating to performance. What Reddy calls "speech acts" I see more comprehensively as bodily practices. Utterances follow conventions or are measured by them. This is how we know, with Smith, when somebody is out of line, transgressing acceptable norms of behavior: the griever who cries too loudly; the socially awkward person who laughs too much; the street-corner evangelist who seems just too angry.

In a hypothetically ideal moral economy, what we feel and what we are expected to emote would accord perfectly. Our utterances would give perfect expression to our inner movements. Reddy asserts that such perfection, such a complete authentication loop of inner states and outward signs, never occurs. Instead, we are involved in a dynamic process of attempting to give utterance to our feelings in accord with expectations, but those expectations in turn affect what we feel. This circular process Reddy labels an "emotive," and the emotive process is never completely successful. The culture in which we are embedded rebounds on us in a physiological sense. Our inner movements are contingent on what we think the world wishes to see of them. In a functional moral economy, feeling states and expression prescriptions will closely accord, and we will derive satisfaction from being able to express emotions that seem to fit how we feel. But considering perfection is not achievable, there will always be a degree to which an emotive fails. After all, an inner physiological state, a *movement*, cannot be projected outward in mere words and gestures in such a way as to perfectly describe it. This is why the emotive process is a form of emotion work, although not conscious.[33] It is an attempt, a striving in Reddy's terms, to express in a way that conforms and, at the same time, to feel in a way that conforms.

Emotive failure can also be severe, and this severe form of failure is a major component of *The Science of Sympathy*. It occurs when the prescriptive codes for emotional utterances are wildly incompatible with feeling states. This can occur on an individual level, but also on a group level. Reddy uses emotive failure to explain the importance of the emotional refuge of the theater in *ancien régime* France, as a place where feeling states could be vented without fear of transgressing strict taboos. Ultimately, when enough people feel at odds with what they are allowed to express, this spills over, outside the refuge, into the real world. Reddy uses the notion of emotive failure to explain historical change. When enough people feel out of place, and when the structures of power that maintain the feeling rules of the old "regime" are sufficiently weak, an emotional—and with it a social and political—revolution can take place.

In my story, there are two principal forms of emotive failure. The Darwinian moral economy of science that sprang up especially after the publication of *Descent of Man,* but incubated since the publication of *Origin of Species,* made severe demands of Darwin's followers, and even more severe demands of his opponents. Even some of the most important proponents of a Darwinian way of seeing the world struggled to reinvent themselves along the lines of new theories of the emotions and the ascription of new values to certain practices. The small nesting doll of the Darwinian moral economy offered radical departures in the meaning and practice of sympathy, but at the same time it had to accommodate people who found it difficult to feel differently, even though they agreed in principle with the new codes of practice. Their emotional lives were, at least as far as their professional activities were concerned, tied to what they did every day. Outside the laboratory, they remained connected to and part of more general emotional and moral concerns. Navigating from the laboratory to the parlor, from the lecture room to the private club, it is important to see that there was no contradiction between their professional and private moral compasses. It has confused many that a vivisectionist could declare himself an animal lover, a proponent of compulsory vaccination could attest to the importance of liberty, or a eugenicist could pronounce the virtues of love. We will not properly understand the late Victorian period until we reconcile such apparent contradictions, while still acknowledging the difficulty of the emotion work involved.

Moreover, an understanding of this emotive process—of trying to feel in a prescribed way by regularly repeating a prescribed expression of that feeling—helps explain why there was such an outcry against the new scientific revolution. If a scientist could only truly feel the new sympathy through a matching process, putting together a complex theory with a highly specialized form of practice, what chance was there for those people who could not or would not entertain such practices, and who would not or could not understand the theories? This occasionally afflicted scientists themselves—Thomas Henry Huxley, for example, could not bring himself to vivisect even though he understood it was moral, and George John Romanes could not bring himself to give up God, even though a material account of nature seemed perfectly sensible—but more often it affected those who were opposed to or frightened of science.

A stark emotive failure leads, I will argue, to emotional crisis.[34] An individual knows what he or she should feel, but finds it impossible to reconcile societal expectations with inner states. Within the Darwinian moral economy, latitude had to be granted to those who needed time and space to work this out. Outside the Darwinian moral economy, those who could not grasp these innovative feeling rules crashed down on the system with reactionary force. The crisis they perceived was much more significant than the emotional hand-wringing of a confused scientist.

These diverse strands of scientific and medical thought and their corresponding practices bring us to a general conclusion. *The Science of Sympathy* is concerned with the dynamic relationship between thinking and doing, feeling and acting. Boiled down to its essence it might go something like this: physiologists cut because cutting did good. It had a high moral purpose. Doing good should feel good. In theory, the act was supremely sympathetic to suffering in the abstract. With practice, cutting felt good, but only because of the emotion work of making an *is* from an *ought*. "This ought to feel right, moral, good," was the refrain, repeated until it did. The practice and the emotional/moral disposition converged on each other to mutually reinforce each other. Though the physiologist projected his argument to the outside world, the real emotion work was being done on himself, in a community—a moral economy—of similar selves.[35]

The same kind of formulation works for the advocate for vaccination, matching medical knowledge to medical practice, and through it finding an emotional outlook on the question of liberty that fits; and for the "priests" of eugenics, a theory of sympathy led to a theory of sexual and emotional practice (eugenics is both) that involved the transformation of the personal feelings and relations of the eugenic scientist. In each case, discrete practices became elements of a whole range of affective practices of the scientific self. The self was defined by this dynamic relationship of feeling/doing or doing/feeling, where practices are affective and affects are practiced. What or who the scientist *is*, to himself and to his peers, is intimately bound with quotidian professional activities, embedded in esoteric intellectual justifications, and constituted through rehearsal, repetition, checking, modification, resistance to difference, and reinforcement by the affirmation of others. Because Darwinism—pluralistic and full of loose ends as it was in its first generation—was a blueprint for living as well as for doing science, its first proponents defined their professions by it as well as being defined by their professions.

Understanding how selves were constructed helps us understand how and why historical actors did the things they did. We can strive to reach fairly intimate appreciations of historical motivations, change over time, and radical differences to the present. To interpret, for example, how Huxley or Galton understood themselves opens a window onto their decisions, morals, actions and, moreover, the decisions, morals, and actions of those who partook of their community, their moral economy. Going yet further, to unravel the complex mesh that makes up a life-world is to demonstrate how difficult it is to engage with that life-world if you are not a part of it. Lives and selves are esoteric. They depend on the experience, education, relationships and everyday circumstance of the individual. If we have nothing in common, we cannot hope, without much work, to be able to access what makes someone else tick. This observation is historically useful because it helps explain historical controversy, dispute, debate, and intellectual and practical incommensurability.

The period covered by *The Science of Sympathy* is defined, by the opponents of Darwin and his putative *ism*, by fear, doubt, and concern that science had taken leave of morality, lost its heart and its sense of compassion, and was taking civilization on a route to an ungodly, materialistic, amoral hell. By telling the story, albeit partially, of how scientists came to think and feel how they did, and do what they did, it becomes readily apparent how difficult it was to communicate the new scientific revolution to the lay public. The distance between scientific selves in the Darwinian idiom and those outside it grew enormously in the last thirty years of the nineteenth century. The esoteric lives of physiologists in their operating rooms, public health officials in their parliamentary offices, and eugenicists in their biometry labs, were increasingly disconnected from, and increasingly unable to communicate with, the nonscientific, nonspecialized general public. This gap accounts for the failure of scientists' simple claims to be "men of feeling," benevolent do-gooders, sympathetic on a high plain. The "science of sympathy" emerged so quickly after the publication of *Descent of Man* that the public was immediately lost, left behind, bewildered by a set of changes and a new type of scientist that they did not, could not, understand.

The book ends with a reflection on the rapidity with which morality can be changed and moral practices *implemented* when specialist communities with sufficient power live out new ideas. The implementation, however, often seems *immoral* to the majority, which has not kept up with the changes or the new knowledge and practices upon which the changes depend. To some extent, the book marks a series of failures, for if a form of sympathy and morality has been bequeathed to us, it is not that of these early Darwinists, whose morality seems so strange. Rather, it is that of the prevailing "common compassion," in its religious idiom, with its roots in Adam Smith. That said, there is reason to be cautious. Did the science of sympathy fail or, realizing its failure to communicate, simply retreat temporarily behind professional walls? After all, vivisection is still practiced, is essentially invisible, and has a profound effect on everything from medicine to cosmetics; smallpox was forced out of existence by the WHO, and vaccines are the stuff of daily news; and eugenic ideas, despite an extraordinarily negative public reception in the twentieth century, live on, repackaged in positive constructions of the prevention of congenital diseases and disabilities. We have, in surprising ways, come to accept the science of sympathy wrought by the first generation of Darwinists. By defamiliarizing sympathy in this historical account, we are challenged to think critically about the meanings of sympathy and morality and the role of science and scientists in our own everyday lives. *The Science of Sympathy* invites reflection on the "evolution as religion" debate that has been central to contemporary critiques of Neo-Darwinistic rhetoric. It charts the speed with which evolution became a new faith, focusing on the explicit statements of its chief apostles. Francis Galton, for example, employed Darwin's own understanding of how religions came about and how they worked to try to wrest control of religion in the name of eugenics and evolution. This argument is developed slowly, throughout the book, but it should become clear by the end that, objections of the Neo-Darwinists notwithstanding, some early Darwinists not only desired the theory of evolution to become a religion, based on their own version of sympathy and morality, but devoutly believed that the health, security, and purity of society absolutely depended on this being brought about. These high priests of science—a clerical identity advocated from within the scientific community—did not themselves need this new faith, but they saw no other way to garner the dutiful support of the masses (who could not be hoped to understand the science). We should also have cause to reflect on historiographical notions of causality for a number of significant developments in the twentieth century, from extreme racial nationalism to the collectivization of state responsibilities. Most significantly, perhaps, the book lays bare the actions of men who, with the conviction that their science was infallible, assumed that they were acting morally. This is a story of things done in the name of being right, notwithstanding that history and science have since proven the "science" of sympathy to be wrong. It is also a story of the serious religious conviction of the rightness of actions carried out with the sanction of science. The science of sympathy, as I will explain here, is a story of *scientism*, or science as a virulently dogmatic creed. As such, it came with a set of taboos, duties, and sins.

2 Sympathy for a Devil's Chaplain

Sympathy and Morality in Darwin's *Descent of Man*

The group of scientific and medical men with whom this book deals took the prevailing understanding of sympathy and changed it. Charles Darwin in particular would lay out an evolutionary account of how sympathy worked to ensure the survival of communities, working together for mutual benefit. Although Darwin's *Origin of Species* did not dwell on the question of morality, Darwin nevertheless pre-empted what was to come in his *Descent of Man*. Natural selection, Darwin argued, would, in "social animals [. . .] adapt the structure of each individual for the benefit of the community; if each in consequence profits by the selected change."[1] Natural selection caused physically "fit" individuals to prosper as parts of a "fit" group. *Descent of Man* would apply this equation to the moral realm, actually serving explicitly to account for the Golden Rule by natural history.[2] Although Darwin fundamentally agreed with Smith and his own contemporary, Alexander Bain, that sympathy was the essential emotional/moral foundation for civilization, he departed from them in a key respect. Sympathy was not acquired mimetically, but was inherited and instinctive. The moral actions that sprang from the feeling of someone else's suffering were not intentional but, in the most civilized communities, automatic.

At the risk of tautology, I want to clarify that Charles Darwin was not a Neo-Darwinist. Given the high degree of Lamarckianism in his works, it is reasonable to claim that Darwin was not even the first Darwinist (that is to say, natural selection was not the only mechanism he thought was responsible for species preservation). The latter point, concerning the inheritance of acquired characteristics, has some bearing on early eugenic theories, to which I will turn in chapter 6. I wish at first to stress the former point, lest there is any confusion. The most important part of *Descent of Man,* in terms of its impact on the moral and social history of Britain in the last quarter of the nineteenth century, would, by the standards of Richard Dawkins, be considered as mere junk. The Neo-Darwinists of the twentieth century, whose hard line has come to embrace the dogmatic mode in the twenty-first, think predominantly of selfish genes. Evolution happens at the level of the individual. Altruism serves the purposes of the individual. There is no such thing, they tell us emphatically, as community, group, or species selection. Evolution does not work for the good of the family (beyond immediate kinship), or of the species, or of society. Whoever says it does, they say, is just plain wrong and goes against "orthodox Darwinian theory."[3]

Darwin said it did. Though in agreement that "altruism" (he did not actually use the word) was self-serving, he said that natural selection worked for the benefit of community.[4] He said it repeatedly. Some contemporaries were critical of this, noting an apparent contradiction between self and community interest. Henry Sidgwick, for example, highlighted the "basic evolutionary reality that it was only the fittest individual who survived," and that what was conducive to the happiness of an individual may have been at odds with what was good for society.[5] But as Darwin pointed out, sympathy was self-serving as well as community-serving, because the most sympathetic individual would likely receive good turns in return for all his own other-oriented action. The society comprised mostly of sympathetic individuals would contain happy individuals and ensure the survival of the community. If the individual sought happiness at odds with that which was good for the group, this was the sign of a less evolved or less civilized individual, not a problem with the theory. If the argument presented here is to be understood, it is key that Darwin's insistence on community selection is taken seriously. It does not matter that Neo-Darwinists disavow this part of Darwin's work. On the contrary, it matters a great deal what happened in the name of this theory being correct, for as long as it was believed to be. Ideas have material consequences. The ideas of influential people have potentially great consequences, socially and practically. It does not matter whether history adjudges those ideas to be true, so long as they were so adjudged at the time they were sent abroad.

Considering that Darwin was not a Neo-Darwinist, and barely a Darwinist, it should not surprise us a great deal that his starting point was not so different from that of the majority of his contemporaries. Yet although Darwin cited Adam Smith's assertion that you cannot have civilized society without the advanced bonds of sympathy among the members of the community—that is, without sympathetic social glue—he and his followers were far from being faithful followers of Smith.[6] In agreeing with Smith about the importance of sympathy, they were also moving away from Smith in attempting to explain the natural history and evolution of sympathy. Their explanations, which also projected into the future, had distinct moral ends. Understanding the nature—"nature" being an operative word—of sympathy was the key to understanding how one *ought* to act in a civilized society. This had implications both at the level of the individual and at the level of the state.

So what did sympathy mean for Darwin and his peers? And what were its implications? Darwin's sympathy bound together family and community units for mutual strength and protection. Without sympathy, society simply could not have evolved.[7] Darwin was responding, in large part, to the first chapter of Adam Smith's *Theory of Moral Sentiments* (1759), viewed through the lens of Alexander Bain's *Mental and Moral Science* (1868), where "the basis of sympathy" lay "in our strong retentiveness of former stages of pain and pleasure." Darwin's reading of Smith was that we feel impelled to relieve the sufferings of others "in order that our own painful feelings may be at the same time relieved. In like manner we are led to participate in the pleasures of others."[8] The sympathizer is likely to reap the benefit of being sympathetic, either in the pleasure of alleviating another's pain or in the reciprocal deeds done unto himself at another stage. Sympathy therefore forecasts—by natural law—the Golden Rule.[9] It is given from one unto another because it would be desired should one be in the other's position. To this extent Darwin was in accord with Smith and Bain, in that sympathy was to some extent "selfish," "for we are led," so Darwin said, "by the hope of receiving good in return to perform acts of sympathetic kindness to others."[10]

But why, Darwin asked, is sympathy "excited in an immeasurably stronger degree, by a beloved, than by an indifferent person?"[11] With all animals, Darwin observed, "sympathy is directed solely towards the members of the same community, and therefore towards known, and more or less beloved members, but not to all the individuals of the same species." This last point caused Darwin some problems. He repeated the point to justify why "savages" were "social animals," despite the fact that "tribes inhabiting adjacent districts are almost always at war with each other," noting that "the social instincts never ['never' bears emphasizing given what follows] extend to all the individuals of the same species."[12] In any case, Darwin satisfied himself with the notion that those communities "which included the greatest number of the most sympathetic members, would flourish best, and rear the greatest number of offspring."[13] With man, sympathy was guided by his "improved intellectual faculties," by "reason and experience" to fashion conduct. This combination of instinct, acquired habit, and reason, produced the *ought* that defined (white male) civilization: "as love, sympathy and self-command become strengthened by habit," Darwin said, "and as the power of reasoning becomes clearer, so that man can value justly the judgments of his fellows, he will feel himself impelled, apart from any transitory pleasure or pain, to certain lines of conduct."[14] It is this civilized distinction that allowed Darwin to retrench a little on his previous assertion that sympathy is directed solely toward the members of the same community. A "savage" he said, would be wholly indifferent about a stranger, but would risk his life to save his kin. A mother would run a high risk for her own child, but not "for a mere fellow-creature." "Nevertheless," Darwin said, "many a civilized man, or even boy, who never before risked his life for another, but full of courage and sympathy, has disregarded the instinct of self-preservation, and plunged at once into a torrent to save a drowning man, though a stranger."[15]

Darwin clarifies that the "moral sense is aboriginally derived from the social instincts." Civilized man judges the "low morality of savages" as being caused by "the confinement of sympathy to the same tribe," and not to the species in general. The civilized man went beyond this low morality by "the simplest reason." The advance in civilization—the uniting of small tribes into larger communities—rendered it patently obvious that man should "extend his social instincts and sympathies to all the members of the same nation, though personally unknown to him." A social instinct is thus rendered entirely subject to a reasoned course of action. The "highest possible stage in moral culture," Darwin wrote, "is when we recognise that we ought to control our thoughts."[16] The simplicity with which Darwin pursues this line of argument is quite breathtaking. "This point being once reached," he said, "there is only an artificial barrier to prevent his sympathies extending to the men of all nations and races." And thereafter, sympathy "beyond the confines of man, that is, humanity to the lower animals" would appear, and indeed Darwin recognized it as one of the "latest moral acquisitions." "At all times throughout the world," he said, "tribes have supplanted other tribes; and as morality is one important element in their success, the standard of morality and the number of well-endowed men will thus everywhere tend to rise and increase." This sympathy at the level of all men, or even all animals, Darwin named "humanity," a most noble "virtue," unknown to ancient Romans and unfelt by savages. It arises as man's sympathies become "more tender and more widely diffused, until they are extended to all sentient beings." A few men—implicitly the civilized, educated, cultivated, white, male men—start it off, and thereafter, he said, "it spreads through instruction and example to the young, and eventually becomes incorporated in public opinion."[17]

Such is the apogee of the natural evolution of morals: a Whiggish, ever-progressing system underwritten by natural law, and led by men just like Darwin. But the progressive cycle is, in *Descent of Man,* suddenly broken by a paradoxical degeneration. It is sympathy that acts as the emotional glue that binds a society. It enables societies to survive, to work together, to shun threats. It is the essential ingredient of civilization. But as the beings in these proto-civilizations evolve toward civilized beings in advanced civilizations, so the power and reach of sympathy grows. And at its highest point, sympathy is extended to the weak, the "inferior races," and even to animals. Henceforth, all the elements of weakness are *preserved* out of sympathy, and the society that has prevailed out of the strength of collaboration has grown up in its own midst the seeds of its degradation. While it sends out the strong and fit and able to fight wars, it tends the feeble, the mentally unstable, and the invalided at home. The essential ingredient of civilization is ultimately also the essential precipitator of decline. I will return to this paradox of degeneration in some detail in chapter 6.

The followers of Darwin took his theory and used it to criticize both the prevailing understanding of sympathy and the moral actions it inspired. "Humanity," for this eclectic group of men, was off course. "Charity" was often misdirected. The sense of someone else's suffering that was inherited and instinctive did not, they variously argued, necessarily lead to the best outcome, either for the sufferer, for themselves, or for the community at large. Beginning with Darwin, these men asked: What if we took sympathy in hand, subjected it to our superior capacity of reasoning, and coupled it to a far-sighted vision of what was actually good for society, instead of merely responding at the instant of emotional reactions to sympathetic stimuli?

Integral to this projection of how society should come to be was a concerted effort to act according to Darwin's image of what the most evolved, or the most civilized, would do. Darwin's theory, and the reformulations of it by others, was implicitly formational of a new concept of self and society, where the acknowledgment of existing weaknesses in the institutions that structured the moral order inevitably suggested new institutions and new structures. Darwin's description of civilization through natural history resolves as a prescription for the natural—controlled, domesticated—future. This prescription was axiomatic: whereas religion had defined duties and morals based ultimately on the fear of, or allegiance to God, promoting societal benefits as a consequence, the new order would define duties and morals based on the good of society itself. The scientist would take the place of the priest, encouraging the lay public to do the best for the community on the basis of facts and figures, and with the fear of committing crimes against the race always in mind. These crimes, unlike sins against religious dogma, had apparently tangible outcomes: drunkenness, criminality, vice, disease. No longer would it serve to say that society's ills were the will of God. On the contrary, society's ills were the fault of social actors who did not know, or who did not heed, the evolutionists' creed. Betrayal of the race would be the new sin. Darwin, gestating his theory of evolution for decades, did not foresee the degree to which this might be radicalized. He simply hoped that the unfit would practice self-restraint. But once his peers started to steer the theory to practical ends, Darwin did become, if not a willing partner, then at least a complicit benefactor of the "new priesthood." The "priesthood" label for these men of science was part scientific conceit and part pejorative name-calling by their opponents. Galton had coined it as early as 1874 in an explicit outline of how his community could and should supplant the clergy. Their "high duties would have reference to the health and well-being of the nation in its broadest sense." Darwin himself had mockingly played with the idea when he passively referred to himself as "a devil's chaplain" to Hooker in 1856.[18] Yet the "priesthood" label was famously used as an attack by Ouida (Maria Louise Ramé) in her 1893 work, *The New Priesthood: A Protest Against Vivisection*, and opponents of Darwinian science blenched when they thought of such an unholy usurpation. The essential point is that the language of religion ran through both the discourse of evolutionary science and the rhetoric of its opponents as each faction staked out the ground of responsibility for moral authority. For the scientists, this was more than useful and convenient analogy. The emotional basis of morality (couched in sympathy) was a commonly held truth. Darwin's appropriation of sympathy, along with his soft-pedalled criticism of the church's now outmoded variety of sympathy, naturally promoted the scientists who understood this to be the forefront of the new moral order.

There is perhaps no finer expression of the dawning realization of the power of Darwinism to *do* or *shape* than R. A. Fisher's publicly proclaimed "Hopes of a Eugenist." Fisher fully appreciated that Darwin's theory was not only "a description of the past" as well as being an "explanation of the present," but also a "veritable key of the future." A thorough application of Darwin's "doctrine" illuminated the implicit connection of biology with morals and emotions, ethics and aesthetics:

> [C]learly we see that not only the organisation and structure of the body, and the cruder physical impulses, but that the whole constitution of our ethical and aesthetic nature, all the refinements of beauty, all the delicacy of our sense of beauty, our moral instincts of obedience and compassion, pity or indignation, our moments of religious awe, or mystical penetration—all have their biological significance, all (from the biological point of view) exist in virtue of their biological significance.[19]

To understand and to be able to fashion biology in its interaction with the world was to have the power to make the world anew. The prescription for a new epoch of progress, with evolution in the hands of men who understood how it worked, was a prescription for the age of the biologist, with his attendants in experimental medicine on the one hand, and in human statistics on the other. It was the very high point of a grand and long-running Enlightenment project to master nature.

Shortly before his death, Charles Darwin set down the story of his diminishing belief in the existence of God, but ended it with an account of how the "wisest men" tend to live their lives. It was to be an expression of the lived experience of a man who had spent decades trying to justify and explain this kind of life by an exploration of evolution according to the laws of nature. A man without belief in "the existence of a personal God" and without hope for a "future existence with retribution and reward" could only "follow those impulses and instincts which are the strongest or which seem to him the best ones," Darwin wrote. Distinct from the more intelligent of animals, like dogs, who acted like this but "blindly," such a man looked "forwards and backwards" in order to compare "his various feelings, desires and recollections":

> He then finds, in accordance with the verdict of all the wisest men that the highest satisfaction is derived from following certain impulses, namely the social instincts. If he acts for the good of others, he will receive the approbation of his fellow men and gain the love of those with whom he lives; and this latter gain undoubtedly is the highest pleasure on this earth. By degrees it will become intolerable to him to obey his sensuous passions rather than his higher impulses, which when rendered habitual may be almost called instincts.[20]

It is a declaration that morality, for the eminent nonbeliever, was driven by emotions that were (or almost were) instinctive. Action was determined by a *feeling*, an "impulse," related to receiving approbation, love, and pleasure. Darwin described a virtuous circle in which acting for the good of others was desirable because it returned good to the source, and this was experienced, ultimately, as "satisfaction." There was no room in a natural historical account of morality for doctrine, commandment, or supernatural consequences. Darwin therefore reduced morality to a naturally derived set of benefits, both personal and communal, that would ultimately aid in the struggle for existence, and that could in turn be consciously directed by the most evolved intellects. Darwin was consistent in this throughout his career.

The autobiographical statement is important because it helps us understand evolutionary scientists as men who sought to live their lives according to the theories they expounded. Scientific expostulation influenced practices of the self. The sympathy that Darwin set in motion impelled evolutionary scientists to mingle theory with ethics and policy, and apply this combination of self, thought, emotion, and ethos to questions of the moral boundaries of individuals, the state, and social institutions. Experience was filtered through, or checked by, an evolutionary account of what experience meant. Darwin and his peers reached new understandings of themselves, their emotions, their actions, and their morals by translating them through a particular construction of how civilization had come to be, and how certain emotions for certain types of human had been instrumental in that process. The capacity to understand this history was predicated on a kind of genius: a highly evolved being capable of rational abstraction and emotional restraint who could fathom the essence of moral action from the tangled history of human development and plot its future course. They strove to *feel* as if they were at the apex of an evolutionary leap. From Darwin himself to his disciples and popularizers (notably Romanes and Huxley), to his sometime antagonists (Wallace), and to those who sought to make change in his name (Galton and Pearson), there were new affective practices in science. The effort, or emotion work, involved in living a science of sympathy would cause a shift in the moral register to be felt throughout the twentieth century.

The Evolutionary Science of Emotions

Charles Darwin's second contribution to evolutionary theories of emotion was published the year after *Descent of Man,* under the title of *The Expression of the Emotions in Man and Animals.*[21] It shares a definition of sympathy with *Descent of Man,* as a "separate or distinct emotion" that directly influences how individuals interact with others in the society around them.[22] But for the most part, the two works do not overlap. Whereas *Descent of Man* focused on applying the theory of natural selection to humans, *Expression* was, in the main, a Lamarckian work that depended on the inheritance of acquired characteristics and habits. *Expression* tied itself in intellectual knots to try to distance itself from the conclusions of Charles Bell, which had dominated the field since the beginning of the nineteenth century. Bell, a noted anatomist and physician, had attempted to capture the emotions of humans and animals according to a strictly scientific anatomical investigation. His self-illustrated deconstruction of the human body's muscular instruments of emotional expression was putatively undertaken for the aid of artists. Bell was dissatisfied with artistic representations of emotional expression, and in particular he took aim at the impossible ideals of classical beauty. By cutting away the skin, the elements of anatomical *design* were revealed, illuminating both the mechanisms and limitations of expression. Armed with such knowledge, artists could better command their brushes.[23]

Yet the influence of this rich atlas of the emotions fell far less upon the world of art than upon the world of science. Bell had provided a schema for decoding human emotional expressions that made explicit claims to universality. Bell's interpretation dominated common scientific understandings of both the origin and, importantly, the meaning of human and animal emotional expressions until the 1870s, when Darwin would contrive ways to revise it. Several key scientific points emerged. First, expressions of emotions in man directly conveyed the "internal emotion" to the countenance, enabling any witness to discern what was going on. Bell did waver on this point, noting that the signs of the passions might only be interpreted according to the experiences that accompany them, rather than being direct embodiments of the passions. But on balance he favored the latter interpretation, noting the "systematic provision for that mode of communication and that natural language, which is to be read in the changes of the countenance; that there is no emotion in the mind of man which has not its appropriate signs; and that there are even muscles in the human face, to which no other use can be assigned, than to serve as the organs of this language."[24] This crucial passage carried an assertion of empirical proof that the human body had been naturally designed—God-given—to communicate the passions of the human soul.

The second point followed naturally from the first. If these expressions were designed to give a sign to the internal passions, then those passions too were universal. Emotions were a "universal language which has been called instinctive," providing humans with an "innate sympathy"—a capacity to understand the emotional lives of other humans—irrespective of "experience or arbitrary custom." The face was an "index of the mind, having expression corresponding with each emotion of the soul." Insofar as the human was a biological universal (and for Bell and his generation, this eliminated what they saw as "inferior" races), so its emotional range and its capacity to communicate inward feelings were also universal.[25]

The third point made by Bell was that this universal was limited to humans and denied to other animals. The "range of expression" in "lower animals" was a "mere accessory to the voluntary or needful actions of the animal." This "accessory expression" was not in "any degree commensurate to the variety and extent of the animal's passions."[26] In other words, what may have been seen in the faces of animals as pain, guilt, love, or pleasure were likely phantasms, or projections of the human imagination. Animal expressions were reduced to the "physical consequences of the necessities of the animal."[27]

To be sure, the more "complex" the animal, the nearer it attained a physical similarity to "man," but the human musculature was unique among animals, being directly related to "his" superior intelligence, a physical sign of "the higher endowments of man, and a demonstration of the peculiar frame and excellence of his nature."[28]

Darwin correlated Bell's analysis, that emotional expressions communicated emotional experiences, with Bell's theology: such an effective mechanism for displaying the passions of the soul must be evidence of intelligent design. Instead, therefore, of explaining how expressions communicated emotions through an account of natural selection, Darwin felt that a rejection of the principle of intelligent design must also entail a rejection of the reason for emotional expressions. Although Darwin clearly subscribed to a universalist position with regard to emotional expressions, he tried to account for those expressions with only scanty references to emotions themselves. Expressions were useful muscular activations, serving some function other than communication, that came habitually to be associated with emotional states. The habits, once acquired, were passed down by inheritance.

Of course, this deep Lamarckian strain in *Expression* did allow for the possibility of historical shifts in emotional expression according to profound changes in culture. But Darwin's analysis in this work always reached for a functional explanation for expressions (which, under the sway of cultural change, could appear in novel circumstances), without getting to the heart of the change in emotions themselves. In the case of the blush, for example, Darwin narrates the close relationship between the attention of the mind and the appearance of the face, and the long-standing exposure of that part of the body to the elements. The blush arose as a function of "self-attention directed to personal appearance, in relation to the appearance of others." Only through the "force of association" did the blush occur in relation to "moral conduct."[29] Darwin described the blush as a human universal that manifested itself in parts of the body that accorded with cultural differences in bodily exposure. To that extent, where or how much a person blushed was influenced by cultural habit. Blushing by dint of moral shame, a phenomenon traceable to a functionally different "exciting cause" would, of course, depend on the construction of morals in a given time and place.

Darwin was ever aware of the shifting ground of cultural prescriptions, even in *Expression,* but gave far less room for the evolution of emotions themselves in that work than he had done in *Descent of Man*. This is perhaps most clearly expressed in Darwin's consideration of "breaches of etiquette" in *Expression*. Etiquette, which Darwin defined as the rules of "conduct in the presence of, or towards others," had "no necessary connection with the moral sense" and was often "meaningless."[30] Yet because the rules of etiquette become fixed—that is, they acquire a moral quality even though they are entirely artificial—then breaches of etiquette can cause shame and blushing. The power of sympathy is "so strong," Darwin stated, that "a sensitive person, as a lady has assured me, will sometimes blush at a flagrant breach of etiquette by a perfect stranger, though the act may in no way concern her."[31] But the evolution and change of sympathy in the course of civilization is largely absent from *Expression*, even if such passages imply it. Indeed, when Darwin considered expressions associated with religious devotion—a combination of reverence and fear—he concluded that because devotional feelings were likely absent from past ages of uncivilized men, the modern signs of devotion such as the "uplifting of the eyes or the joining of the open hands" are neither "innate or truly expressive actions."[32] Where Darwin could not find a functional vestige for an expression he concluded that such an expression was not really worthy of the name.

This leaves open the question of where and when devotional feelings themselves first arose, but those answers were, in many ways, already given in *Descent of Man*, and were reached by a contrasting theoretical model.[33] As Daniel Gross has pointed out, *Expression* in various ways implies that civilization "entails [. . .] a distinct emotional regime that separates it from the uncivilized," a point missed by modern emotions scholars who have pointed to *Expression* as the model for universality *tout court*.[34] Yet the tension between the two works is not broached by Darwin himself and, even though *Expression* sold well, the theoretical distance between the works did not seem to give Darwin's followers too much pause. The immediate influence of *Expression* was not as profound or as wide-ranging as Darwin's treatment of sympathy in *Descent of Man*. The overwhelming reliance in *Expression* on the inheritance of habit, and the profound absence in it of natural selection, make it a strange bedfellow for *Descent of Man* and something of a red herring for scholars of the emotions. After all, one could be forgiven for thinking that Darwin's mature thoughts on emotions might be found in a book carrying that subject in its title. Yet not for nothing is the focus of this work "expression," rather than emotions themselves.

Darwin gathered evidence on the expression of emotions among humans in different cultural settings, asking people to observe subaltern groups and leading them to certain conclusions. Darwin also used the innovative photographic and galvanic techniques of Duchenne to illustrate his point. Duchenne had stimulated, using electrical diodes, the muscles of the human face of a man unable to feel the shocks. He stimulated the muscles to replicate facial expressions that accorded to "basic" emotions, which were captured by photography.[35] Darwin then showed these photographs and others, of actors feigning expressions, to people and asked them to identify the emotions by the expressions. He found a remarkable degree of unanimity among his subjects, and from this he drew the conclusion, though not without circumspection, that humans implicitly understand the emotions of those they see according to the expressive sign that accompanies the emotion.

The conclusion is remarkable for its oversight. No emotion was actually present in the photographed faces. Duchenne observed the produced expressions on the face of his experimental object and decided what emotion the expression represented. All of his emotional categories were preordained. There was no analytical room to say, "I don't know what this looks like." Likewise, Darwin's own research demanded that his respondents tell him which emotion was being depicted. It was highly technical tautology, laced with the irony of being fundamentally a photographic deception. The camera lied, at least about the emotions, from the very first. Duchenne's photographs and Darwin's use of them ought to have suggested how easily deceived a witness might be by a fabrication. Darwin unwittingly assisted in this masquerade, illustrating *Expression* with woodcuts of Duchenne's original photographs, but with the galvanic apparatus and the hands holding it removed. Other illustrations showing muscular formations that are supposed directly to pertain to specific emotional expressions are, with unintended irony, described as being the result of practice and artifice.[36] In one instance, Darwin includes only a portion of a photograph that depicts a woman's attempt to feign grief. The reason given for the crop was that the lower portion of the woman's face showed a different emotion altogether, as a result of "being absorbed in the attempt."[37] The first deliberate attempts to experiment on the nature of human emotional expressions therefore suffered from a central design flaw that went largely unnoticed. The assumption of mechanical objectivity carried enormous influence. The design flaw notwithstanding, this experimental method of manufacturing an expression, devoid of the emotion itself, and presenting it as a "true" example of the emotion, was continually replicated through the twentieth century.[38]

Darwin's attempts, on the one hand, to divert Bell's anatomically based theories into Lamarckian mechanics and, on the other hand, to divert Smith's theory of the sympathetic origins of civilization into a natural-historical account of emotional evolution, were essentially derivative formulations. They were embedded into an emerging science of the mind (and the formal location of the emotions in the mind) to which Darwin himself could only respond. He did not innovate in this field, and indeed struggled to come to terms with what manner of thing "mind" or "intelligence" was.[39] The formation of a psychological science was both rapid and ambitiously wide ranging. The differences and similarities of minds throughout the "chain of being" was on the agenda from the start, and the emotional capacities of animals, the inheritances of Descartes and Bell notwithstanding, were high priorities.[40]

The pioneering work in psychology appeared in the same year as Darwin's *Origin of Species*. Alexander Bain's *Emotions and the Will* (1859) immediately transformed the semantic field that had, until that point, labored with the imprecision of ancient humoral categories. "Emotion," so Bain set out, would from now on be the label to package a whole raft of "feelings, states of feeling, pleasures, pains, passions, sentiments, affections."[41] Absolutely clear at this point in the development of psychology was that emotions prompted action (action was volition), which would become a serious bone of contention later. Bain laid out a physiological understanding of emotion, without reducing "feeling" to "any physical property of matter."[42] This would become another serious bone of contention. For Bain, "the physical fact that accompanies and supports the mental fact, without making or constituting that fact, is an agitation of all those bodily members more immediately allied with the brain by nervous communication."[43] It was a reformulation of Descartes's implicit connection of body machine and mind via animal spirits, but it allowed for greater interpretive flexibility concerning the variety of human emotions.

The advantage here was that Bain could explain apparent emotional differences, especially in terms of expression, between "civilized" and "primitive" cultures. He supplied the first biocultural account of the emotions, dabbling with the effects of what we might call "social construction." The "natural and primitive course of the emotional currents" could be changed through the "power of education," thereby determining an "artificial mode in the spread of the cerebral wave." Bain clearly saw that "natural outbursts" or emotional expressions were "greatly modified by the conventions of civilized man." In the "passion of astonishment," for example, the "open mouth, the sharp cry, the vehement stare, and the toss of the arms" could be substituted by new expressions, "of which a notable feature is the use of artificial language, or speech, for giving the necessary vent to the feeling of the moment." Importantly, the adjustment in expression fed back onto the original state of feeling. Bain effectively described the way in which emotions and their expression work dynamically in the world: "These changes in the allocation of the members that receive the recoil of a state of mental exhilaration have no slight influence in changing the character of the consciousness; for it is not the original stimulus alone, but this, in conjunction with all the reflected waves, that determines the nature of the resulting mental condition."[44] The psychological science of individual emotions was intended from its beginnings to be a *social* science, for humans were directly constituted by the relationship between their nature and their "conventions." To whatever extent Bain could locate "moral sentiments" in the emotional complex of sympathy and "tender emotion," compassion, and "fellow-feeling," he was able also to show that deficiencies in the moral sentiments were compensated by "stimulants from without, namely, the sanctions, or *punishments*, of society."[45] Emotions, insofar as they related to "duty," were therefore always in a process of negotiation with a social body. This was a foundational expression of what would later be developed into Freud's superego. It also meant that Bain could envisage a role for history in order to understand how meanings and understandings of emotional expressions might have come about or changed.[46]

Although Darwin could only react to Bain's work, his chief disciple would try to extend the remit of psychology in evolutionary terms. George John Romanes set out to write a three-volume treatise on mental evolution, in which he tried to demonstrate that the evolutionary recapitulation theory (expounded principally by Ernst Haeckel) worked similarly for minds as it did for bodies.[47] Just as the development of the phenotype rehearsed the evolution of the genotype, so the "psychogenesis" of the human in infancy went through the different stages of evolutionary development attained by other animals. The categorical difference between the minds (and bodies) of humans and other animals, a notion maintained since antiquity, boosted in the eighteenth century by Descartes and again in the nineteenth by Bell, was demolished by Romanes. He insisted that, just as Darwin had shown for the physiological structure of all animals, the difference between the minds of animals and humans was one of degree, not one of kind.

Until relatively recently, Romanes's contribution to the history of evolutionary science and psychology has been largely forgotten. His appraisal of the emotional lives of animals was lost in the wake of opposition to anthropomorphic projections that defined behaviorist science after Morgan's Canon (see below). But in the last decades of the nineteenth century Romanes's work defined the field. His schematic of human mental development went along the following lines: emotions flooded the newly born infant, who at a week old had the capacity of surprise and fear, equivalent to echinodermiata. At three weeks the primary sexual emotions developed, akin to the larvae of insects. At seven weeks (equivalent to mollusks) infants developed parental affections, social feelings, pugnacity, industry, and curiosity, as well as the rudiments of sexual selection. At ten weeks the child was a spider, at twelve a fish, at fourteen a higher crustacean, and by four months a reptile. Along the way, the emotions of jealousy, anger, affection, and playfulness had emerged. Sympathy, the root of civilization, appeared at five months and corresponded (perhaps unsurprisingly) with hymenoptera, which remained the analogy *par excellence* for human civilization. Romanes only followed the emotional development of the child until fifteen months, by which time it would acquire most of the human emotional palette, held in common with apes and dogs (elephants and monkeys knew revenge and rage, but not shame or remorse). Up until that point, these emotions comprised what Romanes termed the "social" emotions. What defined the human in particular came later. Romanes's graph of emotional development had fifty stage points on it. The dog and ape reached only stage twenty-eight. The development of human emotions—"partly human, human, savage, civilised"—all occurred after this point. Unfortunately, Romanes's own death occurred before he could fill in the blanks.[48]

The untimely demise of Romanes left the field clear for new interpretations and directions. Romanes's own disciple, Conwy Lloyd Morgan, oversaw the publication of various manuscripts after Romanes's death, but Morgan took advantage of the opening in the field and redefined it. His famous "canon," published in 1894 (the year of Romanes's death), stated that "In no case is an animal activity to be interpreted in terms of higher psychological processes if it can be fairly interpreted in terms of processes which stand lower in the scale of psychological evolution and development."[49] All of Romanes's anthropomorphic insights were consigned to the scrap heap. Animal emotion, at the turn of the twentieth century, was, if not explicitly denied, then at least labeled different in kind to that of humans. Behaviorism would sound the dominant note for the first decades of the twentieth century.

At the same time, developments in human psychology took a radical turn, with human emotions coming to be defined in a manner completely opposite to that put forward by Alexander Bain. If Bain had drawn attention to the physiological roots of emotional experience, William James would stress an absolute physiological explanation. He too rehearsed the Cartesian formula—mental perception excites mental affection, "called the emotion," and this gives rise to bodily expression—but only to junk it. His completely contrary theory was that "bodily changes follow directly the perception of the exciting fact, and that our feeling of the same changes as they occur IS the emotion." The key passage by James remains a classic piece of mechanistic reasoning:

> Common-sense says, we lose our fortune, are sorry and weep; we meet a bear, are frightened and run; we are insulted by a rival, are angry and strike. The hypothesis here to be defended says that this order of sequence is incorrect, that the one mental state is not immediately induced by the other, that the bodily manifestations must first be interposed between, and that the more rational

> statement is that we feel sorry because we cry, angry because we strike, afraid because we tremble, and not that we cry, strike, or tremble, because we are sorry, angry, or fearful, as the case may be. Without the bodily states following on the perception, the latter would be purely cognitive in form, pale, colorless, destitute of emotional warmth. We might then see the bear, and judge it best to run, receive the insult and deem it right to strike, but we should not actually *feel* afraid or angry.[50]

Expressions of emotions became the emotions themselves, from which feeling states arose. A truly corporally anesthetic person would be emotionless, a "merely cognitive or intellectual form."[51] Crucially, this extreme material view also had the effect of destabilizing emotional categories, since there were presumably an infinite number of reflex expressive actions. James postulated that "there is no limit to the number of possible different emotions" and the search for "true," "natural," or "authentic" emotions was objectively meaningless. Despite his divergence from Alexander Bain, James ended up in the same place, suggesting that we turn to history, in combination with "psychological mechanics," in order to understand *how* any specific emotional expression had come to exist.[52]

Half of James's proposal was enormously influential. He was, in part, responding to a tremendous uptake of physiological experimentation, but his theory of the emotions undoubtedly gave it extra impetus. The infinity of emotional possibilities and the idea of historical change was lost in the quest for the discovery of *basic* emotions, or physiological universality. The innovations brought about by physiological experimentation from the 1860s onward led, according to Otniel Dror, to a "new era in the history of emotions" in which emotion had been "transformed [. . .] into a modern object of knowledge." This object was conceptually different to previous understandings of emotions, but knowledge of this object was also retrieved, stored, and shared in new ways. In the words of Dror:

> The novel interactions between scientific observers, "disinterested" machines, and emotions generated new knowledge and new kinds of relationships: instruments supplanted personal interactions in retrieving intimate emotional knowledge; the interior body superseded language as the primary medium for expressing emotions; the boundaries between private and public, subjectivity and objectivity, and inside and outside were blurred, as experience was exteriorized through laboratory procedures; class, gender, and race were reified through new quantitative emotional measurements and gradations, replacing Victorian qualitative categories; and physiologists and clinicians exhibited and exchanged graphs of their own emotional experiences—producing an economy of emotions inside the laboratory and clinic.[53]

From the blood pressure of frogs to the blush in a rabbit's ears, the emotional state of animals was, with physiological innovation, thought to be presently embodied, recordable, predictable, and modifiable. Any residual doubts about a mysterious substance of mind that might account for the passions was removed by an acute focus on the materiality of affective states. A temperature rise, a quickened heartbeat, a fall in blood pressure: these *were* emotions, findable in the body and its vital organs, traceable on a graph.

This work in mechanically objectifying emotions was complemented by the further use of photography to record the micro-expressions of emotional states. Following the experimental thrust of physiological materialism and positivism, photographic pioneers sought to capture the body's contained emotions, transmitted or represented by the face. Despite the history of photographic manipulation being as old as photography itself, photography came with the seemingly infallible language of *capture*, meaning that the emotional disposition of its gaze could not be disputed. After Duchenne's work had been immortalized by Darwin, others went further, seeking to record the expressions of the mad and the criminal "types," so as to better understand deviance. This science was indispensable to the foundational rationale of Galtonian eugenics.[54] New forms of knowledge would give way to new forms of practice. Sympathy and compassion could be withheld or deployed according to preconceived notions of how expressions of emotion should be received. Putting types of expression through an academic filter, a set of rules of engagement for the sympathizer were newly drawn up. Instinctive reactions of sympathy were likely to be tinged with sentiment, and serve no long-term good. Knowing what emotions were, how they worked, and the good that "true" sympathy could do, armed the self-controlled social and physical scientist with sharper sympathetic tools.

Sympathy and Sentiment

Those scientists were increasingly armed with real sharp tools, instruments not merely of surgery but of experiment. These steel blades would be held up by the men who wielded them as guided agents of moral action; material evidence of the good wrought by the pursuit of knowledge. But the sight of the scalpel, the lancet, as with the hypodermic needle today, could send chills down the spine. The distant effect of improved health, say, is overshadowed by the immediate effect of cutting, pain, and the spilling of blood. The image elicits an aesthetic response. If it is my friend who is to be cut, I might, on seeing the knife, feel sympathy with his forthcoming ordeal. It is precisely this kind of formulation that Darwinian thinking started to unpick.

Sympathy, for Smith, had been an emotional state in itself (the analog of another's emotion) and the basis for moral action (our sense of what is right to do is based on a feeling). No separate concept of altruism had need of explanation or justification. Moral action *was* emotional. Or rather, what impelled Smith to act for the sake of another was the experience of an other-regarding emotion in a context of social conventions. The corresponding emotion in the object of sympathy did not necessarily need to exist. It only had to be perceived. This point was exemplified by David Hume, who argued at length for the emotional basis of morality. Hume made it clear that sympathy was not the direct entering into another's emotions, but rather an awareness of causes and effects. In an observation that is remarkably pertinent to the kinds of context I will explore in the nineteenth century, Hume identified the wellspring of sympathy:

> Were I present at any of the more terrible operations of surgery, 'tis certain that even before it begun, the preparation of the instruments, the laying of the bandages in order, the heating of the irons, with all the signs of anxiety and concern in the patients and assistants, wou'd have a great effect upon my mind, and excite the strongest sentiments of pity and terror. No passion of another discovers itself immediately to the mind. We are only sensible of its causes and effects. From *these* we infer the passion: And consequently *these* give rise to our sympathy.[55]

Smith agreed in almost exactly these terms: the "remote effects" of the "instruments of surgery" might be "agreeable," because they concern the "health of the patient." But since "the immediate effect of them is pain and suffering, the sight of them always displeases us."[56] Context and convention, an understanding of the recipes leading to suffering, gave rise to sympathy. Because the emotions of the other were *not* directly involved, there was always potential for sympathy to be out of place.

Throughout the eighteenth century's coming to terms with sympathy as a basis for moral behavior and as the binding force of civilization, there was a growing awareness of sympathy's uncomfortable relationship with sentimentality. Sympathy as an other-regarding emotion depended, as Smith and Hume were well aware, not on the actual emotions of another but on an imagination of those emotions in the sympathizer. This imagination was based on an individual's own experiences to a certain extent, and the degree to which those experiences could be projected and adapted to form a realization of another's feelings. One consequence of this process was that the other in question did not actually need to feel anything; did not even need to be alive; might have been an animal or an object. Smith described sympathy for the dead, as we might imagine our claustrophobia in being buried and our disgust at being consumed by subterranean "reptiles."[57] This expression of the fear of death demonstrates the human incapacity of imagining the nothingness of corporeal morbidity. When we imagine a body in the ground we imagine it, as it were, as if it were us, alive. Though given by Smith as normal, it is also clearly understood to be irrational.

This irrationality is the basis for a nineteenth-century definition of "sentimentality." It might be defined as sympathy out of place. The concept of justice in sympathy is perverted. The moral dangers of sentimentality increasingly preoccupied nineteenth-century public opinion, to the point where the builders of the science of sympathy decried their opponents as gushing sentimentalists.[58] The knife was just a knife, unless it was the instrument of good. To an experimental physiologist, anyone who perceived horror in the glint of a blade was hysterical. Since the eighteenth century, the sentimental and the feminine had been commonly associated. The fundamental opposition of rational male and emotional female could resolve into a pathology of masculine callousness and female hysteria, labels usually thrown around respectively by one gender at the other and rarely acknowledged among those so labeled. The emotional division between men and women was sharpened by the closer attachment, as the eighteenth century wore on, of the masculine with a countenance of calm resolve, a character marked by its stoicism and a language limited to the blunt, candid, and monosyllabic. This movement away from the "sensible," effete, conversational style of the earlier eighteenth century has been interpreted as a widespread conscious attempt to distinguish British manliness from the effeminate or foppish French man and French tongue, association with which became increasingly politically untenable.[59]

If women were, by popular opinion, emotionally fragile by comparison with men, then the rise of scientific experimentation that was hailed as greatly enhancing understandings of human psychology and physiology only served further to cement gender differences within notions of the natural order of things. Women were blighted by a physiological weakness inherent in having a womb, the functioning of which made them weak and emotional. The ovum itself was, throughout the nineteenth century, considered to be a conservative element in the process of human reproduction, storing the emotional and moral history of the race, but incapable of adding intellectual activity, innovation, or vigor to offspring. These qualities were provided by the male seed, and by extension became male qualities enshrined in nature. A woman who overextended her intellectual activities risked her maternal instinct as well as her fertility. A man who was overly affected by emotional stimuli was likely to underachieve or become socially "dangerous" in his effeminacy. Such rank biologism was a defining motif of the debate around midcentury about whether further and higher education should be extended to women. A vast swathe of public opinion (most of it male, but not all of it) argued that it should not.[60]

The association of the feminine with the sentimental was enhanced by the observably disproportionate involvement of women in a number of reform causes that were frequently bracketed under sentimentalism. The moral reform initiatives of the Clapham Sect, the campaign to end the slave trade, the early stages of the animal welfare movement, for example, were all led by prominent men, but all depended on the activism and financial support of prominent women. From the 1860s, these movements were augmented by the anti-vivisection campaign, the foundations of the National Society for the Prevention of Cruelty to Children, and the beginnings of first-wave feminism. These distinctly moral causes were championed by women as being entirely in keeping with matters of the heart. They were led by "common compassion," an implicit emotional understanding of the injustice of suffering, and the responsibility to try to alleviate it.[61] Opponents of all these causes often based their criticism precisely upon this emotionally driven activism, blaming it for a lack of clear thinking and a lack of farsightedness. These were the charitable excesses of "ladies bountiful," whose insatiable appetite for giving illuminated their lack of understanding of how societal degeneracy was caused and perpetuated. To give aid to all alike went against the design of God, according to certain male theological critics, and against the best hopes for civil society, according to the extreme rationalist view.

Regardless, the remarkable thing about the culture of sentimentalism is that those labeled sentimental refused to acknowledge that there might be anything untoward, misdirected, or misunderstood about their charitable activity. They operated within time-honored traditions of religious charity, the idiom of which extended into proto-humanist activities, and understood their good works as good in the eyes of God, and as having an immediate good effect on the objects of their compassion. Moreover, they rejected the extreme rationalism not only in its criticism of their sentimentalism but also in its own practices of "scientific sympathy." For the believers in common compassion there was no sympathy at all, and therefore no recipe for morality, in the doctrines of science. This is more fully explained in the following chapter. Yet for a new generation of scientists this wave of sentimentalism was only an indication of the inferior emotional development of the mass of people.

Susan Lanzoni proclaims that, for the last few decades of the nineteenth century, "sympathy reigned supreme in the Anglo-American intellectual world as a vital social emotion, one that had organic or physiological elements, but that also extended into the realms of ethics and aesthetics."[62] There was something self-congratulatory in this, for sympathy "marked a higher evolutionary stage of sociality, and was present to a greater degree in the civilized as compared to the 'lower races.'"[63] The immediate problem facing evolutionists who subscribed to such a view was, to them all too apparent, the relative unevolved nature of, frankly, most people. The scientists themselves, the visionary priests of the new order, implicitly marked themselves out as the only living embodiments of the highest, if not the next, stage of evolution.[64] Describing how the process of evolution had worked, and forecasting how it would work, they put themselves at the forefront of evolutionary advancement. Because they were the first to see the damaging consequences of the tendency for sympathy to be self-serving and to understand the importance of community selection, they necessarily saw themselves as the instruments of change at the level of public opinion and education. Society was on a natural evolutionary course laden with suffering and pain. This path could be sped up, with a diminution of suffering, if only society would be led by those who understood how to play nature's game, artificially selecting over a short period what nature would select over eons. Scientists justified themselves as the natural leaders of society, and demonstrated the incapacity of everyone else to form sound judgments about the progress of the race, by laying out an elaborate history of the evolution of emotions. The most refined emotions, those properly engaged with advanced capacities of reason, were to be found in the bodies of eminent European men. After that a sliding scale of emotional frailty was marked out according to sex, class, age, and species. The opinions of the emotionally frail, however forcefully voiced, could therefore easily be ignored.

What emotions were, from a scientific point of view, was privileged knowledge. Vernacular assumptions about passions and sentiments allowed for the ready misunderstanding of the essence of emotions. By appropriating this knowledge as the purview of science in an exclusive manner, scientists of emotion were able to imbue scientific discourse with a naturalistic narrative that ordered the world in the way they expected or desired to see it. Essential to this *ad hoc* appropriation of knowledge of emotions was the perhaps unconscious project of putting scientists themselves at the apex of evolutionary development. Crowning themselves as the first to have reached a true understanding of the nature of emotions and their function in society, they inevitably saw themselves as the masters of their own emotions, whereas others remained slaves to instinct and ingrained habit. To understand the mechanism of the evolution of emotions was effectively to demonstrate the ways in which their further development could be "domesticated," that is, taken in hand, subjected to an academic, laboratory-based, statistical method of manipulating, reprogramming, and remaking of emotional habits and emotional inheritance. The science of emotions in the age of Darwin went hand in hand with the remaking of the human itself. Coming from various disciplines—physiology, biology, psychology, statistics—scientists were able to mobilize a new politics of sympathy and transform it—reify it—into concrete new social practices, replete with their own moral justifications.

Herbert Spencer thought that the sympathy Darwin described would continue to evolve with the continuing enhancement of human intellect to ensure that society would not break. Humanity would be better equipped to see the long-term consequences of its actions, and to decide on the best overall moral action. Spencer made clear the new distinction in types of sympathy. Put succinctly, the more evolved the emotional being, the more considered, and the less impulsive, would be the conduct of that being. It would be better equipped to feel the long-term consequences of its actions, and to decide on the best overall moral action. "An emotional nature not well developed," Spencer said, "will be relatively impulsive—the liability will be for each passion to display itself quickly and strongly, without check from the rest." With a higher development of the emotions, "there will be little liability to sudden outbursts of feeling." The resulting conduct, derived from a more complex and "a greater number of feelings severally less excited," was likely to be "more persistent." This checking process, slowing down sentimental or aesthetic reactions, meant that the morality of an action could be correctly weighted. Spencer, of course, was outlining the contrast between civilized and "savage," but, as was typical, he averred that an illustration of his theory was "furnished by the contrast between men and women."[65] If a continued evolution of the emotions was to lead civilization into a higher moral culture, the science suggested that elite men would be at the vanguard of progress.

Taken together, evolutionary conceptions of the natural history of morality were designed as description, but ended up as prescription. They were motivated by a need to understand how civilization and its moral code had come to be, without resorting to theistic interpretations. Indeed, theism had to be explained too. It is difficult in the writings of Darwin in particular to find any manifesto or agenda for social change. Darwin espoused the values of his time, and understood the importance of the institution of the Church and of the ways in which society was stratified, by class, race, and gender. He sought primarily to explain how this had come to be, which meant ascribing an enormous amount of importance to the role of religion. Religion provided the ethical basis for community, institutionalizing the Golden Rule (which could be explained by natural selection) and building an authoritative agency through which "the fit" could broadcast the values that would ensure the survival of society. Public opinion, influenced by the more evolved, dragged along the less evolved in such a way as to "domesticate" nature. Religious institutions had mastered the brutal nature of human animals and turned them into natures that could be guided, controlled, and reproduced, instilling a sense of duty. Evolutionary scientists would inherit this mantle. The demise of religion, with the replacement of cleric with scientist, was all a part of the natural development of things.

It was normal, of course, for Darwin's critics to overlook the evolutionary importance he ascribed to religion (in the main because his description of it seemed to be based on a materialism that ultimately signaled religion's end). One of the more spirited accounts of the evolution of society that accepted in general terms the principles of evolution, but defended religion and predicted its continued importance, was written by Benjamin Kidd. Kidd's substantial intervention was to insist that religion not only accounted for the extraordinary social cohesion that had allowed Western Civilization to flourish, but also that this was never a rational intervention and could not be supplanted by a rational turn. Channeling William Lecky, Kidd asserted that the religious governance of mankind had been facilitated by speaking "to the heart," not to the mind.[66] The purely rational evolution of humanity had tended in the direction of purely selfish and presentist interests, and whatever the evolutionists predicted for the future of humanity, no amount of rational advancement would end in serving the community.

Kidd's work is useful because of the mistake it makes in the reading of those texts that explained the evolution of society. Kidd could not attribute a cause to the rise of religion. Religious formation was part of natural law. He largely neglected Darwin's own words on the subject, choosing to focus instead on criticisms of Spencer and Huxley. But *Descent of Man* might have given him pause in his assertion that community cohesion was impossible without the supernatural. Darwin, for his part, ascribes the rise of religion to the sympathetic instinct. As sympathy develops, so humanity outgrows religion. While Darwin would have agreed with Kidd that the development of major religions was not conscious or rational, his prescription for the future evolution of humanity was that the sympathetic instinct become hitched to reason. There was no age of pure reason predicted in the pages of Darwin. On the contrary, the advance of mind was to be aided and abetted by the advance of emotions. By being able to see the structural benefits of social cooperation wrought by religious organization, Darwin foresaw the capacity for humans to transcend religion's institutional constraints and pursue the benefits unhindered. Darwin was not, however, celebrating religion, and even though he viewed his own civilization as the pinnacle of evolutionary forces to that date, his own theory necessarily predicted further change and improvement. The sympathy of the Judeo-Christian tradition was, in Spencerian terms, ultimately ego-centered.[67] Whereas Smith had seen this and trumpeted its virtue, Darwin and his peers saw the limitations, at best, and the dangers, at worst, of a self-serving philosophy (even where the outcome was apparently altruistic). Having extolled the virtues of religion's socially cohesive history, Darwin then necessarily pointed out how it was now doing more harm than good.

The Scientific Man of Feeling

Religion had been central to Darwin's history of the evolution of civilization. It had institutionalized the most advanced instincts of sympathy and broadcast them to those who might not have acquired them "naturally." The impact of public opinion was key not only to the story of how Victorian civilization had come to be globally predominant, but also to the future development of humanity. The old priesthood had presided over the Golden Rule, but the "new priesthood" would usher in a new order of sympathetic judgment and its moral correlates. Darwin and his contemporaries preserved a label of eighteenth-century coinage to make the transition seem both more subtle and safe in their hands: the "man of feeling." What had been a badge of openness to tenderness, sensibility, even sentimentality, in the eighteenth century, was reinvigorated in the late nineteenth, partly as a defensive strategy in the face of a storm of concern that the "new priesthood" was completely without feeling and partly as strategy of positive reinforcement of new practices. The public needed to be persuaded, but men of science also had to reassure themselves. They felt moved to advert to their capacity for feeling, but in language modified to fit the new circumstances.

The finest expression of how the scientific man of feeling had to operate—literally—was given by William Osler in 1889. Osler, who is celebrated and claimed by Canada, the United States, and England as a preeminent physician, was significant as a public defender of vivisection on both sides of the Atlantic.[68] In 1889, before a class of new graduates in medicine at the University of Pennsylvania, Osler delivered a speech on the importance of "equanimity," or the embodiment of self-control and emotional control during and through medical practice. It is a blueprint text for the history of emotions, for it demonstrates an explicit awareness of both the need and the capacity to change one's feelings according to one's context, and to work, literally, through feelings. It is a statement of the importance of deliberately bounded and planned affective practices, of science and of the self. Osler knew that the qualities of the "imperturbable" surgeon were kindred with the laboratory physiologist, and saw a need for a selective deafness toward the tidal wave of sentimentality that seemed to threaten progress in the operating theater and in the laboratory.[69] The practitioner would be lost if he felt his patient's pain. He urged his new young colleagues to have their "nerves well in hand" and to avoid the slightest facial expression of anxiety or fear even under "the most serious circumstances." To fail in this regard betrayed an inability to put one's "medullary centres under the highest control," and would lead to disaster. "Imperturbability" was a "bodily endowment" that ensured "coolness," "calmness" and "clearness of judgment in moments of grave peril." It was character defined by "phlegm":

> Now a certain measure of insensibility is not only an advantage, but a positive necessity in the exercise of a calm judgment, and in carrying out delicate operations. Keen sensibility is doubtless a virtue of high order, when it does not interfere with steadiness of hand or coolness of nerve; but for the practitioner in his working-day world, a callousness which thinks only of the good to be effected, and goes ahead regardless of smaller considerations, is the preferable quality.

He urged his young charges to "cultivate [. . .] such a judicious measure of obtuseness" that would "meet the exigencies of practice with firmness and courage, without, at the same time, hardening 'the human heart by which we live.' "[70] To the outside world, their practices would simply appear to be callous, but they would know the higher plane upon which they were operating. Their apparent callousness was for the greater good. This argument defined all the various discourses of a science of sympathy throughout the last quarter of the nineteenth century and into the first decade of the twentieth. It echoes Spencer's exhortation to resist outbursts of feeling, to tolerate a present evil for the sake of a distant good. New research, it was argued, was necessary if new methods and means for the alleviation of suffering were to be reached. If that research was unpleasant, difficult, or disgusting, then so much the better if it could be done by men who could overcome these difficulties.

Physiologists were such men, in Osler's opinion. They had a quality of the "experimental spirit in medicine," with which there was "nothing else in human endeavour to compare from the standpoint of humanity." He agreed with his colleague Harvey Cushing that there was a "feeling of regret [. . .] that animals, particularly dogs, should thus be subjected to operations, even though the object be a most desirable one and accomplished without the infliction of pain," but his conclusion was clear: the "humanity of the physiologists" could be trusted implicitly. This humanity—compassion in the broadest sense—had been adhered to through "lives of devotion and self-sacrifice," through a useful callousness, and carried to an "incalculable" extent.[71]

"Humanity" was a functional synonym of compassion in the nineteenth century, and it was claimed by the scientific and medical community in competition with a number of societies and special-interest groups that staked out new claims for the meaning of "humane," especially with regard to the treatment of animals. It was important to point out, as frequently as could be managed, that opposition to new scientific endeavors was based on a false "humanity," compared to the true humanity of science. When George John Romanes reviewed a book that defended vivisection, he drew attention to the anonymous author's "large and generous heart" and his "finely strung feelings." This was a lover of animals as well as a "lover of men."[72] Romanes was Darwin's chief disciple in the 1870s and was among the last of the old school of gentlemen scientist amateurs. A man of independent means, he was a scientific generalist, largely eschewing academic bowers, and doing much of his experimental work at home. His family and closest friends called him "The Philosopher," a classical reference to his all-rounder abilities, but also to his "sweetness of temper and calmness of manner," according to his wife.[73] Social graces went hand in hand with what were increasingly becoming professional practices. Romanes loathed the demise of the former in favor of the latter, dismissing mere professionals as "paper philosophers," arrivistes who risked losing their hearts for the sake of knowledge. As Jim Endersby has identified, the ideal gentlemen scientist pursued knowledge disinterestedly and independently. To be labeled "philosophical" rather than "professional" was "the highest accolade one could bestow on a scientific man."[74] As such, Romanes lived and practiced in this philosophical idiom, and looked askance at anyone who impugned his character or the character of his class. "[O]ur physiologists as a class," he told Frances Power Cobbe in a public letter in *The Times*, "are not less English gentlemen because they are highly cultured men of science." On the contrary, these experts were uniquely positioned to see the value of physiology, both as scientific progress and as public good. As such, they had sought "legislative restriction" of vivisection in order to dispel "suspicions cast upon themselves by the inflamed imaginations of unscrupulous agitators."[75] Physiologists, so the claim went, were animal lovers *and* beneficent actors through their vivisection-based research. Only those who knew nothing of the methods and ends of vivisection could, Romanes argued, see this as a contradiction in terms.

Romanes also defended his mentor, John Burdon-Sanderson, against antivivisectionist attack by calling attention to his "purest humanity."[76] For his part, Burdon-Sanderson also understood why the opponents of science, who were typically female or, if male, painted as effeminate, could not comprehend the higher states of feeling of men of science. They were "handicapped by emotion" by "virtue of their [physiological] organization." Women were more likely to have their "bodily functions" disturbed by emotions, leading them to deranged actions and opinions unfiltered by rational judgment. It meant that they "will always fail in the race," but it gave scientists a justification for being misunderstood and further motivation to continue to assert their special breed of tenderness.[77] As Romanes put it, the "mental chariot" of women was easily overturned, leading them into false positions and illogical politics.[78] By way of contrast, the emotional chariot of elite men was in check. For the new man of science, the ideal was for feelings to be either filtered through knowledge, subjected to a will conditioned by scientific wisdom, or for feelings to be conditioned by repetitive practice, where scientific work was also emotion work. Projecting their minds forward to the abstracted humanitarian aims of reducing suffering in the world, they worked without immediate concern, unperturbed by the immediacy of blood, pain, or acrimony. *Doing* science would be an affect in itself, the sign of which was the apparent absence of affective behavior. Just as with the rise of observational objectivity, unveiled as an affect by Daston and Galison, so the even-tempered, apparently "unfeeling" operator was defined and activated by emotional means and affective practices.[79]

The scientific man of feeling, equipped with his self-defined superior intellect, emotional disposition, social insight, enhanced understanding of what was "good," and new experimental means to effect change, felt understandably compelled to act. His recipe knowledge had been enhanced. To have had the insight to understand how civilized society had come to be, and to project where it might go, accorded to him the status of society's natural leader. He was an embodiment of the evolved future he projected for everyone else. That future, a future of structurally reduced suffering on a grand scale, depended on the advance of medical science, public health, and the further "domestication" of the "race." Because the experimental science on which these things depended could only be carried out by men of such refined sensibility, who understood how sympathy really worked, they felt it was left to them to bring about the improved future.

3 Common Compassion and the Mad Scientist

Transnational Controversy

The controversy over vivisection began in earnest in 1873. There had been a major expansion of physiology in continental Europe, and the first signs of a significant institutionalization of physiological practices in England were clearly emerging. In part, this depended on a Darwinian cosmology, whereby advanced physiological understanding of animals would illuminate the physiological understanding of man. Contiguity among species was key. There had been rumblings of public dissent in the 1860s, about which more below, but the real spark for public outrage came with the publication of the *Handbook for the Physiological Laboratory* in 1873.[1] The specter of young novices cutting up animals in their private rooms led to a yellow-press campaign that spread fear and impugned these men of science who seemed impervious to the infliction of pain. Were they completely without sympathy? Public attention was focused by a Royal Commission on the Practice of Subjecting Live Animals to Experiments for Scientific Purposes, followed by the Cruelty to Animals Act of 1876, by which animal experimentation became subject to a government licensing system. Seeing the dangers of the storm of public opinion, prominent figures within the scientific community had themselves lobbied for licensing in order to allay fears and to regulate the use of anesthetics. The public inquiry of the mid-1870s encompassed the following questions: the utility of experimental research; the "humanity" of physiologists at home and abroad; and the degree to which animals could, or should, suffer pain. Within medical science in general, there was little dissention with regard to the benefits already derived, and the wealth of humanitarian relief to follow, from physiological research. The difficulty lay in the moral price at which those benefits were purchased.

The dominant note of concern was not for the plight of animals, or for the morality of mad scientists, but for the character of the nation. England, so the argument went, was under threat from the importation of Continental, specifically German, scientific methods that paid little or no regard to suffering. German scientists were painted as blunt instruments of a *profession*, doing the bidding of the state without question and without feeling any responsibility to consider the sensibilities of the public. English scientists, on the contrary, were still touched by the essence of the amateur generalist. This was changing rapidly, causing moral panic among interested observers, with science being chopped up into salaried specialisms, detached from any public accountability. The alarm bells were sounded ten years before the publication of the *Handbook*, with the key rhetorical devices being introduced.

Moritz Schiff was the scapegoat, introduced to the English public by dispatches from Frances Power Cobbe, writing from Florence.[2] Cobbe, about whom more below, was disgusted by this hard-hearted German physiologist who was exploiting a general lack of regard for animals in Catholic Italy to carry out a routine schedule of torture on dogs. Cobbe's dispatches lay dormant for a decade, until revived by Richard Hutton in the *Spectator*. Hutton would also become a leading anti-vivisectionist voice, motivated by an acute concern that the feeling of tender mercy was dying. I deal with the roots of that concern later, but for now it suffices to recount Hutton's description of Schiff's work, and his warning to English scientists. Hutton had no truck with German *Gefühl*, dismissing it as having "no kindred with genuine feeling."[3] The Germans were without the capacity for genuine sympathy. Schiff "did not scruple to make Florence ring with the scream of his living subjects," Hutton reported, with the rider that this should sound a "warning to English physiologists of the loathsome insensibility to which the habit of vivisection is apt to lead." Schiff, to Hutton, represented "the most hideous depth of human hard-heartedness."[4] This German disposition was at the border, threatening English hearts with contamination.

Skepticism about Continental callousness was the loudest-sounding note in the Royal Commission of 1876, fanning the flames of xenophobia. The moral collapse of the character of the English scientist, in thrall to the German methods they had learned firsthand across the German-speaking continent, was only the thin end of the wedge. Scientific publications aimed at introducing or popularizing new specialisms to young neophytes were thought to carry a risk of contamination. If the young hearts of the nation were blunted from the moment they entered the ranks of science, and if these men went on to be society's leaders, public voice, and conscience, then civilization itself was at risk. At least, so went the criticisms.

The *Handbook*, compiled by John Burdon-Sanderson and some of his most eminent colleagues, was just such a threatening publication, claiming at once to be an introductory guide to physiology, but without stressing the need for anesthetics and seemingly without any cautionary advice about the need for expert supervision. Sanderson himself came to regret those omissions as his character was called into question in front of the Commission of Inquiry. Hutton, who had done more than most to bring the character of scientists into question, was appointed to serve the commission. He became the principal agitator against the leading voices of physiology. Hutton asked Sanderson if he wished to see "the education here more like the type of education in Germany," baiting him by blurring the lines between pedagogy and moral character.[5] He did the same to others, suggesting that Germans succumbed to a greater "zeal" in the pursuit of science, allowing them to override a natural tenderness "to pain," which was likely to catch on among a generation of new English specialists, all of whom had been trained in Germany.[6] Hutton found an ally in the Royal Society for the Prevention of Cruelty to Animals, whose representative told the commission at Hutton's asking that he feared the importation of "Continental usages [. . .] into this country."[7] He also found support among the older generation of anatomists and pathologists, who had little if any experience of German procedures or characters. Do not "our own medical men [. . .] come back with these new methods in their heads[?]," he pressed. Does not the atmosphere in Germany have a "natural effect" on those Englishmen who study there?[8] Henry Acland, Regius Professor of Medicine at Oxford and president of the General Medical Council, supplied Hutton with the words he wanted to hear. "[T]hings were done [. . .] habitually" in Continental laboratories that "would not be tolerated" in England. The Germans went about things with "unscientific carelessness" that would be "hurtful to the moral sense of England" if they were allowed to penetrate the security of the then present wave of "public feeling [. . .] against the infliction of purposeless and uncalled for suffering upon living animals."[9] The concern was that this bulwark of public feeling was being eroded by the scientists themselves, who did not keep their work behind closed doors, but rather broadcast it to general audiences, in handbooks like Burdon-Sanderson's and in public lectures at which polite society met.

The frailty of this public sensibility to the expanding practices of physiologists seemed to be demonstrated by some of the testimony given to the

Royal Commission. It looked as if all the fears of Continental influences had already come true in the person of Emanuel Edward Klein.[10] Klein was Slavonian, of Austrian descent, and had trained in Vienna and the laboratories of Ernst von Brücke and Salomon Stricker. From 1871 Klein found himself in London, working at the Brown Institute under John Simon and, notably, John Burdon-Sanderson. Sanderson had already defended himself against the charge of introducing not only German methods but a German lack of sensibility, and John Simon had testified that Klein had not suffered any "hardening effects at all with regard to sympathies with the lower animals."[11] Klein was to make all that sound like so much eyewash. He claimed on public record that he had "No regard at all" for the sufferings of animals, and that he only used anesthetics for convenience's sake, to hold the animal still. As a teacher, he used anesthetics only because he considered the "feelings and opinions of those people" who bore witness. Otherwise, he repeatedly attested his complete indifference to "the sufferings of the animal," extending that principle to all the physiologists of Europe, and opining that there was no difference in feeling "amongst the physiologists" in England. Klein only lamented the degree of public interference. In Europe, physiologists were left alone to get on with the job. If only it were like that in England, Klein implied. Alerted to the seismic effect of his testimony by his colleagues, Klein later tried to undo the damage by retracting his testimony, claiming poor command of English. The commission eventually printed his original testimony and his revised testimony, with the latter in the appendix of the report. It made little difference. Despite Klein's claims that the suffering he caused was so minor as to be negligible, and that he was "as much opposed as anyone in this country to unnecessary or unprofitable cruelty to animals," the die had been cast. He noted bitterly that the English public differed from the European in its disposition "to take care of other people's consciences in matters they do not clearly understand."[12] That note went unheeded by almost everyone, but it actually summed up the argument of physiologists and evolutionary scientists at large. They were dealing with new knowledge and new understandings, new recipe knowledge for new ethical and affective practices. One could only comprehend it, feel it, live it, from the inside. Klein was there; Hutton and his ilk were not. There was no use explaining it, for it could not be explained.

The Scientist Monster

Klein looked for all the world like an archetype of the mad scientist. Dim fears of Continental monsters looking west became acute once the public became aware that those monsters were already among them. No wonder the self-proclaimed scientific man of feeling felt such a need to advertise his professed tender sensibilities. Klein's original testimony pinched nerves of self-doubt that everyday scientific practices were, in fact, harming scientists and harming society. A long-running fear of becoming callous prompted doctors, surgeons, and scientists (especially anatomists) to run self-tests on their sympathetic feelings. At the core of Neo-Platonic and Enlightenment thought was the notion that cruelty rendered its perpetrator cold. The boy who tortured animals grew into the cold-hearted murderer, according to William Hogarth's classic formulation of the "Four Stages of Cruelty." The murderer in that sequence ends up on the anatomist's slab, but the anatomists themselves are implicated in this "reward" of cruelty. The matter-of-fact dissection of the human flesh—the body an object of study, without sanctity or reverence—coupled with an unfazed attitude to the sight of blood and the stench of death, were themselves signs of diminished tenderness, or we might call it compassion fatigue. This was a long-standing belief. Smith, for example, observed that "One who has been witness to a dozen dissections, and as many amputations, sees, ever after, all operations of this kind with great indifference, and often with perfect insensibility."[13] Men who saw human bodies as instruments of knowledge might lose sight of the limits of what was acceptable in their discipline. How long before the need for knowledge caused anatomists artificially to swell the numbers of dead for their own ends? The Burke and Hare scandal seemed portentous of moral collapse, but even without running to such lengths, the anatomist's regular dealings with body snatchers and grave diggers suggested a loss of essential human dignity, both for the scientist and for the dead. What if they shifted their gaze to the living?[14]

That had long been the plight of the surgeon. Inured to the sight of blood and benumbed to the suffering of others, the surgeon inflicted pain as a matter of necessity, at least until the practice of administering anesthetics became widespread in the second half of the nineteenth century (see chapter 4). The advent of anesthesia had a material effect on the capacity of surgeons to operate with Osler's required equanimity, but in the popular imagination surgery, and the figure of the surgeon, remained specters of pain and fear. Moreover, the figure of the operator remained shrouded in suspicion. Would not such a man be incapable of sympathizing with the ordinary plight of the pained and diseased? Would not his methods justify or inculcate a disregard for suffering in general, so long as the perceived end of it could be deemed worthwhile?

Such were the known emotional risks of the profession, but it was not until the 1870s and 1880s that, claims about being men of feeling notwithstanding, the medical profession embraced its equanimity, its calculated callousness. William Osler's speech is exemplary in this, but the principle was widely adopted and accepted by scientists in each of the fields I cover in this book. A small amount of suffering could be justified to prevent a larger amount. And to inflict this small amount of suffering, practitioners needed to cultivate a procedural callousness, being mindful only of the ends and not the means. They had to exercise an emotional control—to suppress their instinctive reactions—to stretch their feelings farther and do what they had determined to be real good in the world. The knowledge of this greater good would preserve their tender sympathies, rendering them not only safe but also a positive benefit for society.

Nevertheless, with men like Osler prescribing that sympathetic instincts be turned off, subjected to a new form of scientific rationalism that Darwin and company had made available, it is hardly surprising that moral traditionalists started to ring the alarm bells of fear. For those outside the Darwinian narrative, that is, engaging with it only in the form of reaction and conservative defensiveness, it seemed as if Darwinism justified the procedural infliction of pain and suffering, enabling the "fit" to survive and prosper and condemning the "weak" to a cruel fate. It was not an enormous stretch to imagine what might happen if the compassionate instincts of scientists were switched off permanently. The emergence in popular literature of the mad-scientist trope had a tremendous effect in stoking the fires of fear about the new laboratory experimentalists. For people who had never set foot in a laboratory or an operating room, this literary window onto the hidden chambers of science was at once illuminating and terrifying. The scientist of the Darwinian stamp lurked in a gruesome theater of horrors, morally stunted, emotionally cold, and a dangerous virtuoso of cruel arts.

Oscar Wilde's 1891 novel, *The Picture of Dorian Gray,* contains an exchange about sympathy that gets right to the heart of common fears of the encroachment of a highly literate and scientifically expert callousness: "'I can sympathise with everything, except suffering,' said Lord Henry, shrugging his shoulders. 'I cannot sympathise with that. It is too ugly, too horrible, too distressing. There is something terribly morbid in the modern sympathy with pain. One should sympathise with the colour, the beauty, the joy of life. The less said about life's sores the better.'"[15] Such was the shocking amorality of the aesthete, seeking out the visceral pleasures of all life had to offer, at the expense of mere sentiment. Experience was all. Talk of entering into the pains of others is cast as a distinctly modern phenomenon. Had not the pains of ages past been simply accepted? Disease, injury, and death simply occurred. Part and parcel of the hand dealt by life, there was little point in dwelling on the plight on the countless thousands who succumbed to smallpox, for example. Such was one interpretation of the ethical shelter of the past. Engaging in sympathy with the masses was, for Lord Henry, a futile exercise, but his coldness did not go without reproach:

> "Still, the East End is a very important problem," remarked Sir Thomas, with a grave shake of the head.
>
> "Quite so," answered the young lord. "It is the problem of slavery, and we try to solve it by amusing the slaves."
>
> The politician looked at him keenly. "What change do you propose, then?" he asked.
>
> Lord Henry laughed. "I don't desire to change anything in England except the weather," he answered. "I am quite content with philosophic contemplation. But, as the nineteenth century has gone bankrupt through an over-expenditure of sympathy, I would suggest that we should appeal to Science to put us straight. The advantage of the emotions is that they lead us astray, and the advantage of Science is that it is not emotional."[16]

The passage is a lucid elaboration of the idea put forward by Haskell, discussed in the introduction to this book, that a general lack of concern for the plight of the working poor in England was caused by the availability of an ethical shelter that enabled the privileged to go about their lives without the slightest ethical concern about the slavery on their own doorsteps. No ready solution—no recipe knowledge—spurred them into action. But whereas real scientists were working out new recipe knowledge, new everyday practices, that would compel them to tackle the plight of the suffering poor in new ways, tearing down the ethical shelter of the likes of Lord Henry, Wilde took the opportunity to implicate science in the further removal of ethical concern. Rather than having a grand vision of societal suffering, science is painted as being completely devoid of emotions, its practitioners coolly putting things in order. Wilde captures the essence of the central critique of science in the age of Darwin, which focused principally on the means of scientific research—emotionally restrained and detached procedures—and seldom on the ends. The better world envisaged by *fin-de-siècle* scientists is veiled by the preoccupation with quotidian brutality.

Wilde was conjuring with a popular trope, that science and scientists were killing the compassionate hearts of the English in the name of the acquisition of knowledge and, like Lord Henry, they actively despised the common displays of sympathy of the well-meaning patrons of charities and good causes. The principles of scientific research, and particularly of scientific experiment, were frequently criticized for being amoral. In the hands of eminent men whose voices tended to carry influence in the shaping of public opinion, to be cold and amoral was to open the door to a dangerously *im*moral view of society. That so many of these men advocated experimentation on living animals, the imposition of medical research on an unwilling public, and the elimination of the "weak" members of society who were thought to be the harbingers of racial degeneration, reinforced that implication. With such men, so the common representation went, charity would go to the wall; suffering would be rendered nothing more than an abstract concept; and progress would be reduced to the accrual of facts.

It is no surprise that in the hands of literary magicians, the latent fears of a morally disjointed scientific priesthood could be whipped up into stories of demonic genius and unscrupulous inhumanity. Wilde's Lord Henry, the Mephistophelean tempter of Dorian Gray, speaks the voice of science without being party to it. Nevertheless, the high praise of science as unemotional and blind to suffering is fittingly and typically delivered from the mouth of the Devil.

The passage above informs the novel's dramatic climax, as Dorian Gray calculates the best way to dispose of the murdered body of Basil Hallward. The emotionless scientist has a combination of peculiar qualities that enable him both to be blackmailed and to be undaunted by the sight of blood. Alan Campbell, who had "spent a great deal of his time working in the laboratory" at Cambridge, was enough of a gentleman to feel the pinch of shame with which Dorian threatened him, but enough of a scientist to be able to carry out the task of reducing a body to dust without being moved. Though there is clearly a great moral gulf between the scientist and the errant aesthete, there is no disagreement between them about the effect of scientific practice on the scientist. "You go to hospitals and dead-houses," Dorian tells Alan, "and the horrors that you do there don't affect you. If in some hideous dissecting-room or fetid laboratory you found this man lying on a leaden table with red gutters scooped out in it for the blood to flow through, you would simply look upon him as an admirable subject. You would not turn a hair. You would not believe you were doing anything wrong." Dorian alludes precisely to the cold reasoning of the beneficent scientist of the day, whose work was always a means to an end: "you would probably feel that you were benefiting the human race, or increasing the sum of knowledge in the world, or gratifying intellectual curiosity."[17] But in unmasking Alan Campbell, Dorian demonstrates that this claim to a greater good has nevertheless come at the price of permanent emotional damage. Whereas Dorian the murderer aesthete cannot face the gruesome task of disposing of the dead, he assumes, correctly, that the scientist will be unmoved by the procedure. Ultimately, facing the ruin of his reputation as a man, Alan the scientist enters the room with the dead body. Dorian, the killer, shudders and says he cannot go in, but the cool scientist, resigned to his task, simply states that "It is nothing to me," and sets about the instrumentalization of morbid flesh.[18] We discover later that the threat of shame has led to his suicide. Meanwhile Dorian gets away with murder.

If Alan Campbell's scientific attitude ultimately played a part in his undoing, Wilde at least painted that character with the verisimilitude of Victorian social mores. Nevertheless, it is the practice of science itself that allows for the compromising circumstances, because faced with a scene of extraordinary disgust, Alan is predictably unmoved. He is, in this sense, as monstrous as Dorian himself. When taken to extremes, this monstrosity becomes nightmarish. H. G. Wells's 1896 characterization of the mad scientist was the notorious Dr. Moreau, a Gower Street physiologist in exile, conducting crude vivisections to make monstrous beings. Removed entirely from the "ordinary" world, detached from the niceties of domestic, public, and gentlemanly life, Moreau can no longer define the complexities of "humanity," whether his own or whether as justification for his actions. Rather, he declares, "so long as visible or audible pain turns you sick [. . .] you are an animal," later adding: "Sympathetic pain—all I know of it I remember as a thing I used to suffer from years ago."[19] These are signs, conveyed with Wells's irony, of the brute. The assertion here is that, without sympathy, without the capacity to be sickened by the sight or sound of pain, a man becomes a mere creature, denuded of its human—humane—cachet of distinction. The end that Moreau has in mind is an advance in knowledge and technique. Any means are justified in the name of the acquisition of knowledge, even if they cost a man his soul. Moreau is a monster who gains his verisimilitude from his rehearsal of contemporary physiology's well-intentioned justifications. Physiologists often pointed to the advances in knowledge gained by their experiments, but knowledge itself was to be instrumentalized into practical medical, surgical, and scientific gains. Immediate signs of pain could be shrugged off in the name of ultimately saving a greater pain. Yes, sympathetic pain in the laboratory was shut off, deferred in the name of sympathy for humanity, but this was not at the expense of sympathy for pain in general or in the abstract. Concern for suffering at large was a leitmotiv of physiology in these years. We shall have cause to reflect upon these justifications at some length in this book. Moreau mastered the first parts of the equation, switching off sympathy, but failed to retain a grip on the moral purpose of this emotional control. This makes him monstrous, and he is literally isolated, forced from civilization, his name a scandal on Gower Street (the real-life professional base of Burdon-Sanderson, Klein, and others).

The archetypes of the mad scientist do not antedate Moreau by much, although Wells capitalized on the contemporary controversy aroused by vivisection to magnify the plausible horror of his tale. Stevenson's *Strange Case of Dr. Jekyll and Mr. Hyde* (1886) set the standard for narratives of the degeneration of the scientific self. Jekyll's loss of control of his monster alter-ego was caused, it should not be forgotten, by the experimental excesses of the otherwise moral and civilized Jekyll himself. Science itself is the agent of dehumanization. This, of course, nods to Shelley's *Frankenstein* (1818), in which the process of creating a person in the laboratory leads to the double failure of the emergence of something not human *enough* in the form of the monster, and of the moral and mental decline of Dr. Frankenstein himself.[20] Nineteenth-century literary representations of the scientist commonly cautioned against the personal, social, and moral pitfalls of unchecked experimental ambition. We have inherited the mad scientist as a horrific figure, recognized by his all too obvious criminal excesses. In the nineteenth century, the horror of this figure lay in the prospect of his actual emergence. We have retained the notable literary works, but lost the discursive context in which the callousness of science and of scientists was on the lips of the conservative establishment, in the leading articles of journals for the well-heeled, broadcast from pulpits up and down the land, and chewed over by the guardians of the ivory tower, that bastion of character building.

The instrumentalization of the evil scientists for propaganda purposes, on both sides of the Atlantic, was startling. By the time the anti-vivisection and anti-vaccination controversies hit Germany and the United States in the 1880s, they brought with them from England an already matured discourse of scientific monstrosity and developed it into a frightening account of the reality of the mad scientist. The English anti-vivisectionist journal the *Zoophilist* carried a regular column, totting up the "hecatombs" of Louis Pasteur, representing him as the age's most notorious mass murderer. Similar representations were made in the analogous journals in the United States and Germany. In the American *Our Dumb Animals*, the scientist teacher, "Satanus Inferno," replete with serpent's tail, holds aloft a live cat and a large knife. In front of him, a crowd of children rejoices in their instruction, themselves wielding knives and cats, one of them hoisting aloft the severed head of his cat on the point of his knife (fig. 3.1).

The association of emotional blunting, moral corruption, and youth was a common theme. Physiological science was shown not as the loss of innocence but as the corruption of innocence. It was as if moral boundaries were not merely transgressed, but rather had never been acquired. In another image in *Life*, a group of physiologists stand around an operating table, about to perform a vivisection on a large dog. At the door, a young boy asks: "Please mister! Have you seen our dog?" Physiologists were frequently accused of stealing household pets to supply the raw materials for their cruel designs (fig. 3.2). The encounter is far-fetched, but struck a chord with the public. The

Fig. 3.1: "The Great Educator," *Our Dumb Animals* (1897)

Fig. 3.2: "Please mister! Have you seen our dog?" *Life* (1910)

Fig. 3.3: "The inoculation maniac," *Anti-vivisection Review* (1911)

moment captures the callousness of the scientists as well as the corruption of the child. These scientists do not need a demonic countenance in order to be depicted as having gone beyond the pale with regard to morality.

The fear of these experimentalists went beyond vivisection per se. As indicated by the sustained attack on Pasteur's toxicological programs of research, the anti-science lobby was horrified by all scientific activities that used animals in research, convinced that such objectification stemmed from a lack of tenderness. Moreover, lasting fears of being contaminated with animal matter and animal diseases made it easy to paint medical scientists in particular as the harbingers of hybridity, or demonic chimeras. Relying on the power of the state to enforce his designs, the "inoculation maniac" would force the unwilling to submit to his animal poisons in the name of health care (fig. 3.3).

For many in the West, such interventions were horrifying and heretical acts. The human frame was sacrosanct, and the appearance of disease was a manifestation of God's will. To vaccinate was to risk bodily pollution and the contravention of the divine order of things. That discourse had been around since the first vaccinations at the end of the eighteenth century, but despite the enormous success of vaccination against smallpox, the tide in favor of vaccination was turning as the nineteenth century went on. Much of this will be dealt with in chapter 5, but suffice it to say for the present that the expansion of immunology, from Jenner's first vaccine to concerted efforts to counteract anthrax, rabies, and diphtheria in particular, were met by large sections of the public by genuine concern about the moral and emotional cost of this new medical "progress." The increasingly strict enforcement of vaccination against smallpox, especially in England, was further cause for concern. The state had entered into the private domain of the family and, for the first time, was telling parents what to do with their children. To the fearful it appeared as if monstrous scientific innovation had the backing of the highest worldly authority, but that conscience seemed to have been lost in the bargain.

In order to understand what was at stake with new scientific innovations and practices, one must grasp the great public level of concern about the moral risks of doing science, for the scientist, and of science being done, for the public at large. This discursive context took the form of, on the one hand, a sustained attack on the personalities and institutions of science as they emerged from midcentury, and on the other hand, a sustained defense of prevailing moral codes that depended on a commonly received notion of sympathy and compassion. This analysis of the "common sense" of sympathy is essential if we are fully to grasp the magnitude of the intellectual and practical departure made by those who would profess a new sympathy in the name of Darwin and natural law. This *mise-en-scène* of sympathy had long roots in the eighteenth century, about which scholars have spilled a lot of ink. But it was declared most loudly and most vociferously in the mid-Victorian period, when the cracks began to open up.

Common Compassion

What, according to popular opinion, was wrong with the hearts of scientists? I want to lay out the forcefulness of the specific moral and emotional reactions to Darwinism, which border occasionally on the theological, as the best way to understand prevailing notions of compassion and its moral consequences. The outcry at Darwin's theory often took the form of an objection to the moral gaps it could not fill, as well as to its apparent threat to traditional Christian outlooks, not at the level of divinity but at the level of everyday emotional practice.

There was no greater or louder critic of Darwin outside the world of science than Frances Power Cobbe. She was a noted Unitarian writer and campaigner, especially remembered for her vociferous attacks on vivisection under the auspices of the Victoria Street Society, which she founded in 1875 and which worked for the complete prohibition of the practice of experimentation on living animals. Cobbe wrote principally about public morality and the consequences of immorality, cutting her teeth writing correspondence for the *Daily News* from her base in Florence. She had single-footedly kick-started the anti-vivisection campaign in England by complaining about a German (Moritz Schiff) experimenting on dogs in Italy. While her outrage at vivisection has often been (mis)interpreted as being borne on the wave of early animal rights activism, Cobbe's particular outrage is better located among concerns about the state of civilization as it reeled from Darwin's interventions. It was Darwin's *Descent of Man* in particular, which Darwin had personally given to Miss Cobbe, that fired up her sense of imminent moral—and therefore mortal—danger.

Cobbe's understanding of sympathy and compassion, pity and tenderness, benevolence, mercy, and humanity—all of which were roughly synonymous in her conceptual world—was inherited from Smith and Hume, and particularly from Smith's interpretation of the Golden Rule. It also depended on a much older tradition that exhorted the civilized to be humane to those beneath them in order that their habits be kept pure. A heart hardened in the chastisement of children, dumb animals, or the poor was thought to lead with certainty to callousness in behavior to one's peers, friends, and family. This process of heart hardening, with its roots in Neo-Platonic philosophy, the teachings of Kant, and the popular prints of Hogarth, was still in common usage in the late nineteenth century. But just as it was commonly understood that the sowing of cruelty to animals in childhood led to a harvest of murder amid a calcified justice system, reaped in adulthood, so the converse was equally true: to enter into sympathy with lowly objects helped cultivate it between humans. In Kant's terms, animals were analogs of humans. Morality should be practiced on them, as an indirect duty to humankind, lest brutal deeds taint the tender heart.[21] Cobbe hoped that the cultivation of "pity and tenderness to the beasts and birds" would cause "the hearts of men [to] grow more tender to their own kind."[22] Moreover, Cobbe seemed well aware that to sympathize with an other, particularly an animal, was not to feel its emotions but to understand the images of cause and effect. Just as Hume's pity was aroused by the preparation of the instruments of surgery, so Cobbe's was aroused by the "sight of suffering."[23]

In an argument about the necessity of removing slaughterhouses and cattle markets from the center of the metropolis, Cobbe made a case not for animal rights or for the suppression of a carnivorous diet but rather for the control of sights and sounds liable to cause society to grow callous. There was a "*ricochet*" of man's cruelty to animals, a collateral injury that few had perceived. "The sight of suffering which we do nothing to relieve, has in itself the most hardening effect, till the time comes when the instinct of civilized men to pity and relieve dies out altogether."[24] In effect, her solution for Hume's problem of seeing the instruments of surgery was to lock such instruments away, to keep them from the view of all but those qualified to use them. The public's heart—its tenderness—depended on being noncognizant of the recipes of suffering. The capacity to sympathize was limited. Too much exposure to suffering, and the heart would become inured to it.

The problem with science, particularly with the rapidly professionalizing set of evolutionary disciples, was that they had precisely opposite motivations. They sought to spread their gospel far and wide, publishing in polite and general journals, writing letters to the principal newspapers, giving public lectures and courses in prominent places, and generally advertising their creed wherever they could. That creed, as it turned out, included the medical and societal benefits wrought from physiology's *modus operandi*, vivisection. Here then was the direct challenge to Hume's sensibility and to Cobbe's need to keep the dark places of what she called cruelty out of the light. To harp continually on the benefits to society of medical experimentation was to present to the public the image of the scalpel, the operating table, and the bloody hand, again and again. Whatever strides forward this might have indicated for science, it came at the cost of the public's emotional condition.

Since scientists were themselves great authorities in the arena of public opinion, and since they seemed to have no compunction in sharing their methods and results with the lay world, Cobbe felt she had little choice but to shut them down. With an unintended irony, she chose to risk shining light in those dark places in order to create a scandal and an outrage, all the time stating her view that the moral sense depended, as it had always done, on the divine, not on the bald stuff of natural history.[25] Both Darwin and Cobbe worked with an understanding of the "moral sense." Moral action was inspired by moral feeling. Darwin, according to Cobbe, had accounted for the moral sense as "an instinct in favour of the social virtues which has grown up in mankind," but that "would have grown up in any animal similarly endowed and situated." So far as it went, and pertaining only to that "glimmering of something resembling our moral sense often observable in brutes," Cobbe gave Darwin's theory its due. But she thought the overall picture of "the moral nature of man" unsustainable. The "conception of the principle of Benevolence," which owed its capital B to its divine origin, could not possibly be extended to the conduct of worker bees to drones. Likewise, it was surely a "relief" that the "destructive work" of animals was done "guiltlessly" and "devoid of moral sense." To think otherwise, or worse, to act as if Darwin had written the truth, was horrifying.[26]

By 1885 it seemed to Cobbe that Darwinism had ushered in the blueprint of a "New Morality," which, "if ever generally adopted, would 'sound the knell of the virtue of mankind.'" The first chapter of this "New Gospel of Science," as she phrased it, was: "Blessed are the merciless, for they shall obtain useful knowledge." She reduced the Darwinian scheme to the following rubric: "the argument amounts to this: Nature is extremely cruel, but we cannot do better than follow Nature; and the law of the 'Survival of the Fittest,' applied to human agency, implies the absolute right of the Strong (*i.e.*, those who can prove themselves 'Fittest') to sacrifice the Weak and Unfit *ad libitum*." There were degrees of affective attachment to this dogma, according to Cobbe. The "most candid and honest adherents" of the new maxim upheld their views as a "corollary from the new natural philosophy," a bowing down to "Nature's plan." Since "Man" was "merely a part of Nature," what better than to simply "fall in" with it. But infinitely more shocking to Cobbe were those who "proceed to argue that in following Nature's apparent recklessness in inflicting suffering, man will be *obeying God*." Such an argument obliged each individual in her own way to "carry out those tremendous and pitiless laws which govern the hurricane and the earthquake, and which are exemplified in the instincts of the vulture and the tiger."

Such references to the divine were, in the name of science, nothing but disingenuous appealing in Cobbe's eyes. "The truth is," she wrote, "that these modern Men of (merely physical) Science are so absorbed in their material researches that they have actually dropped out of sight all the moral and spiritual sciences together; and they go about in the footsteps of Mr. Darwin, endeavouring to gather the grapes of Morality off the thorns of Physics and Zoology." To Cobbe such hopes of a moral harvest were forlorn, and the attempt to deduce "some rule of duty" from the study of nature was an indication of the "childishness of science."[27] The future of civilization hinged on the question of whether "we shall hold [. . .] by the morals of Christianity or by the morals of Darwinism; by the belief that the voice of the conscience is the voice of God in the soul; or that it is the tune played by our music-box brains, merely as a result of the 'set' they have acquired from the prejudices of our ancestors.'" That equation hinged on which version of sympathy was adopted as the guiding principle. For

Cobbe, the principle of sympathy was that "The Strong shall bear the burden of the Weak." She interpreted Darwinian morals, hitched to the notion of the survival of the fittest, to indicate that "the Strong shall inherit the earth and the Weak be trampled in the dust."[28]

Vivisection was therefore the principal cause for concern among the scientific practices that followed on from the "new morality." Here was a practice that seemed to institutionalize cruelty and to justify it in terms of knowledge gained. The Darwinian disciple was not the "low ruffian," consumed by wayward passions for blood and death, but the "calm, cool, deliberate" scientist, "perfectly cognizant of what he is doing; understanding, as indeed no other man understands, the full meaning and extent of the waves and spasms of agony he deliberately creates."[29] Those men, in a position of great influence, posed a great risk. As such, one of the Victoria Street Society's explicitly expressed aims was to preserve the prevailing tender sympathy of the Christian, as defined by Cobbe. "Our object," Cobbe wrote in the Victoria Street Society's official organ, the *Zoophilist,* "would preserve the whole community for the present and all coming generations from the deadliest possible injury, namely, the suppression of compassion, and the fostering of selfishness and cruelty, in the high places of education from whence those vices must permeate the whole character of the nation."[30]

Physiology, as we shall see in chapter 4, defended itself almost in the terms Cobbe supplied: as a calm, deliberate practice, but one aimed at raising compassion to a level the likes of which Cobbe could not conceive. But the burden of proof seemed to lie with science, since Cobbe spoke to a lay audience in terms of "common" decency and an emotional life that she presumed every Christian to understand. Upon reading a magazine spread in the *Illustrated London News* in 1884, including a picture of a "rabid 'inoculated dog in a cage,' and of a whole room-full of rabbits waiting the same doom" in the laboratory of Louis Pasteur, Cobbe wondered whether "science teaching has curarized our hearts that they can beat no more with any natural emotion of indignation or pity." Cobbe subsumed Pasteur's toxicological investigations under the head of vivisection, which became a general term for all experiments involving animals, even though it technically only should have concerned experiments that required cutting. Thus the toxicological work of Pasteur, Lister, and their peers in the search for vaccines, anti-venoms, and medicinal sera, all came under the same scrutiny as causes of sympathy's decline. This potential for societal emotional calcification, brought about by the "New Benefactor of Humanity," as Cobbe sarcastically labeled Pasteur and company, was a direct result of a lack of mercy in high places. "May God have mercy on His poor creatures," Cobbe said, "for Science has none."[31] In a famous open letter to Charles Darwin in *The Times* at the height of the vivisection controversy, Cobbe asked whether "the principles of the evolution philosophy require us to believe that the advancement of the 'noble science' of physiology is so supreme an object of human effort that the corresponding retreat and disappearance of the sentiment of compassion and sympathy must be accounted as of no consequence in the balance?" The unavoidable cost of physiology in the name of Darwin was surely too high a price to pay. "What shall it profit a man," Cobbe asked, "if he gain the whole world of knowledge and lose his own heart and his own conscience?"[32] Cobbe's influence went far beyond English public opinion. She was directly influential of the German anti-vivisection movement, and found an echo in the activism of the founder of the American Anti-Vivisection Society, in 1883, Caroline Earl White. She, for example, denounced Alexis Carrel of the Rockefeller Institute in New York as a "thinking machine," whose "finer instincts of humanity are drowned in the one passion for so-called scientific research."[33] Emotions in physiologists were essentially dead.

If Cobbe is the best example of an individual with a sustained attack on the moral economy of Darwinian science, her writings nevertheless exemplify a general concern about the state of public morals. Where science after Darwin was criticized, it often involved a fear of increasing callousness at the expense of tender emotions: pity, mercy, sympathy. Where Cobbe appealed to a particular group of politically motivated women and their male supporters, others who voiced similar concerns reached a wider audience and deepened the impact of the argument. It is important to recognize that the rhetorical mobilization of the phrase "common compassion," or its analogous constructions, did not come out of nowhere. They were based on a long-standing appreciation of how society worked, how bodies in a commonwealth related to each other through sympathetic feeling. Thus, the Royal Commission on Vivisection concluded its report in reassuring tones. Not only was the commission reassured about the "humane spirit" found in some foreign laboratories, but also the "general sentiment of humanity" at home continued to "pervade all classes in this country."[34] Hutton, who refused to sign off on such unbridled optimism, wrote his own minority report. In it, he noted the deplorable decline of "common compassion," which had collided with "the pursuit of scientific truth." For him, "the ends of civilisation, no less of morality" required that this common compassion be followed. To be inured to pain and/or to the sight of blood could never serve civilization, since the signs of brutality would, through the ordinary functioning of public opinion, catch on, just as common compassion would catch on if demonstrated from on high.[35] Later, as Hutton resumed his anti-vivisectionist campaign in the *Spectator* and elsewhere in the press, he continued to lament this rise of the unfeeling scientist. A small amount of "carelessness and callousness," when "justified and even extolled by men of high standing," was "far more prolific of moral danger to the character of society at large" than the same sentiments among "rude persons."[36] With gentlemanliness came public responsibility. Scientists, claiming title to gentlemanliness, were failing in their duty of right feeling. With their blunted emotions in the name of a false good, they pointed society in the wrong direction. Civilization was about to be pricked by thorns in the name of pursuing the fruits of knowledge.

Alexander Bain's psychological studies seemed to provide support for such an assertion. According to his *Emotions and the Will,* compassion, or pity, was "sympathy with pain" and a "sure source of good actions under all forms of distress." The "sympathizer takes on the pain that he witnesses" and "works it out as if it were his own." This mixture of "tender feeling" and "sympathy" is activated through "fellow-feeling" for slaves, victims of domestic abuse, subjects of tyranny, those in ill health, and those who succumb to "the tragic incidents of human life," as well as "when a man is deprived of his rightful earnings, by force or by fraud" and under all other circumstances of injustice.[37] In sum, this common compassion defined the Victorian social conscience, and was the motivating force behind Victorian charitable initiatives of all kinds. The age of utilitarianism was about the happiness of the greatest number, but utility depended on an understanding of the nature of suffering: on the capacity to suffer *with*. In a society where evil was defined by suffering, pleasure in alleviating it was also led by the mutual feeling of pain in the form of compassion. The Victorians were aware of the paradox, even if historians have not been. "The giver of aid derives the direct pleasure of pity," said Bain, who also called "pity" a "luxury." Moreover, this luxurious pain was necessary, for if "the feelings of compassion, sympathy, and dutiful respect" were suspended, then, according to Bain, this would "leave the field free to the other passions": The "protection that habitually surrounds a man, without which he might be at any time a victim of the sport of every other man" is removed.[38] To be without compassion would leave one callous, impervious to the surrounding suffering. Should men of influence be callous, then society as a whole would be risked.

This, in short, was the fear of those who saw in the rise of Darwinism after 1859, and even more so after 1871, a bleak materialism, subject to an amoral natural law in which the fittest would survive and the strong would crush the weak. To be a proponent of Darwinism, and/or its attendant practices, it was averred, was to be a man without feeling, and a man with no basis for morality.[39]

4 Sympathy as Callousness? Physiology and Vivisection

The State of the Heart

This chapter is concerned with the new sympathetic practices of physiologists, the ways in which they attempted (and often succeeded) to control their emotions in their everyday work, feeling for the suffering of others through a suspension of feeling in the immediacy of the moment and the setting. The strategies they used to implement vivisection as an affective practice of Darwinian sympathy were many and varied. I deal with them in turn, carefully demonstrating the ways in which numerous discursive threads were intertwined with quotidian procedure, a meticulously designed array of materials, and the pseudo-domestic ambience of the laboratory itself. We begin with the defensive claim that physiology in England was morally safe because of the distinct set of emotional qualities that inhered in the very essence of being English. From there we move to the banality of the everyday in laboratory design and laboratory procedure, before focusing specifically on the emotional benefits of the use of anesthetics in particular. Anesthetics proved to be not only politically important in safeguarding physiologists from the charge of the procedural infliction of pain on the innocent, but also emotionally important in allowing them to act on benumbed objects that, when considered in the abstract, did not have any pain to be reflected in the sympathetic hearts of scientists. The chapter concludes with some reflections on the difficulties that evolutionary scientists might have had when attempting to practice what they preached. A theoretical guideline for a higher form of sympathy was all very well, but what if the looked-for equanimity was always supplanted by a shaky hand and a sick feeling in the pit of the stomach?

Though the new band of specialist physiologists admitted their indebtedness to German methods and insights, they rejected the charge that they were Germanizing science and society, making callous what was once tender, by adverting to their uncorrupted and reassuring Englishness. National character in this case was constructed as a set of emotional and moral qualities: a bundle of virtues that fitted out the English scientist uniquely to go about his business and retain his sense of humanity. The modern man of science might have to perform operations distasteful to the layman, but his scientific practice did not, so he would argue, make him any the less of an English gentleman.

George John Romanes, as we have already seen, had put it in precisely these terms, but he was only developing and echoing much of what had been rehearsed in defense of vivisection in the testimony given to the Royal Commission in 1876, which in turn had looked to the defense of Moritz Schiff, the first target of anti-Continental science. E. Ray Lankester had taken up the baton on that occasion, insisting that physiologists had "not become callous to the sufferings of animals."[1] Ray Lankester spoke from within the inner circle of Darwinian science. He had been introduced to Darwin, Huxley, and others when a child, and he had taken the opportunity of the Radcliffe traveling fellowship to study physiology at Leipzig and Vienna, morphology with Ernst Haeckel at Jena, and marine zoology with Anton Dohrn in Naples.[2] He was an exemplar of the young Turk physiologist, seeking out Continental knowledge and practices, but housing them in a particularly English character. He defended Schiff as "a humane and kind-hearted man" and pointed out the degree to which the physiologist himself "often suffers most acutely from his sympathy with the animal, but controls his emotion and endures his pain in companionship with the dumb animal for the sake of science."[3]

This sense of "companionship" was already on the way to being an English cliché, but physiology tested the limits of an assumed quality of national character that cared for the welfare of animals. After all, pitching vivisection as animal welfare was always going to be a hard sell. Nevertheless, the fears of continental contamination notwithstanding, the claims of a particularly English sensibility came from on high. Michael Foster, a co-compiler of Burdon-Sanderson's *Handbook* and first Regius Professor of Physiology at Cambridge, declared a faith in the durability of "the good feeling which characterizes the Englishmen who have taken part" in Continental methods.[4] That sentiment was echoed by most of the young Turks. F. W. Pavy of Guy's Hospital, who had witnessed the experiments of the "father of physiology," Claude Bernard, in the 1850s, averred that English students were too sensitive to withstand exhibitions of callousness. In his program of cutting living animals, he assured his students that "no experiment will be introduced which will wound the feelings of the most sensitive" among *them,* the students. Preserving the heart was paramount.[5] Others, notably Dr. Rutherford of Edinburgh, Dr. McDonnell of Dublin, and Sir William Gull, agreed: "anything like cruelty or indifference to suffering would be scouted by the public opinion of the students."[6] Burdon-Sanderson himself, who had many laudatory things to say about the work done on the Continent, was keen to point out the essential difference of the English heart. The things that were done abroad that "ought not to be done on humanitarian grounds" would not be done in England because of the "quite different" "sentiment" of "physiological workers" in that country. To be sure, he was confident too of the "reasonable humanity" of the "leading men in Germany," but left an impression that English hearts calculated all the better.[7]

Those who had had extensive experience in honing their skills with German masters concurred. William Rutherford, who had learned his craft in Berlin, Dresden, Prague, Vienna, and Leipzig, was full of praise for men like Carl Ludwig, but knew that the "English race" had a higher "tone of feeling" than the Germans, and employed a greater "amount of delicacy" in the institutions of medicine. English physiologists, he thought, were rather "more careful about repeating a painful experiment" than the Germans, and the Germans also went "rather further" than an Englishman might in his tuition of students. If there was an apparent indifference to suffering in Germany, the English would be safeguarded from it by a difference in "national temperament."[8] Experience in the physiological laboratories of the Continent remained essential in an English physiologist's training, but so long as they adapted German affective practices into English ones, there was no danger of moral corruption.

The formal defense of vivisection was therefore built upon a combination of strategic and affective lines of argument. There was a laudatory survey of the state of the art: physiological science was furthering knowledge, improving the skill and fidelity of its practitioners, and reducing suffering in the world. There was also a positive appraisal of the state of the heart: the skill and fidelity of those practitioners was shot through with tender feelings, a conscious but controlled sympathy, with an eye firmly fixed on the mores of gentlemanly conduct. And there was an appeal to the heart of the state: this was England, not the Continent. Insofar as physiology had made itself available to the scrutiny of the state, offering itself up to a system of licensing to safeguard the public from harm or fear of harm, it had chosen to partner with government as a security against cruelty and against callousness. Darwin himself had been at the center of that initiative. With government co-opted into new physiological research, the state itself bought in to the notion of physiology as a sympathetic practice, signaling its public benefits for the present and for the future. As discussed in chapter 2, scientific men assumed that their powers of abstract reasoning, combined with a certain doggedness that enabled them to endure the tedium of experimental methods, equipped them to be the exemplars of a most highly evolved capacity of measured emotional judgment.

Physiological Theaters of Emotional Control

The physiological laboratory was the most contested and controversial site in the scientific world in the last third of the nineteenth century. It was styled by scientists as the site of medical progress, the epicenter of civilization's quest against the ravages of disease, and the place at which gathered the foremost men of knowledge and skill, who together would lead society out of darkness. To physiology's opponents, the laboratory was nothing but a glorified torture chamber. Such are the stakes in the history of physiology and its associated spaces.[9] If our societies are the inheritors of this debate, it is the latter strain that springs most readily to mind. The "ghastly kitchen" of the vivisector, ruled over by the mad or monstrously inhumane scientist: This is the stuff of literary cliché, B-movie horror, and the anti-vivisectionist street activist.[10] This imagery is also well represented in the historiography, by historians who have found it difficult to imagine how to reconcile the man of feeling with the practice of vivisection. The mad scientist prefigures, even contaminates, empirical research. Patrizia Guarnieri observed: "On the one hand, the white-collared scientist who tied down an etherised dog on the operating table who [. . .] opened its skull and removed the cranial lobes. On the other, the gentleman who always had some delicacy in his pockets for the animals, and made sure that they lacked neither food nor affection. A sort of Dr. Jekyll and Mr. Hyde perhaps."[11] She is not the only one to have employed literary fantasy as analysis. Stewart Richards critiqued the physiologists of the 1870s and '80s thus: "Whatever their ethical imperatives as private citizens (when they were evidently no less humane than other men), they were able as professional scientists, temporarily but repeatedly, to suspend 'normal' sensibilities in a way that we may recognize as more widely familiar throughout history than the singular case of Dr. Jekyll and Mr. Hyde." He went on to wonder whether John Burdon-Sanderson, one of the pioneers of physiology in England, had fallen, "like Dr. Moreau [. . .] under the spell of research," which was the "source of a psychological commitment to specific instrumental norms that overwhelmed or obscured any more broadly based ethical misgivings."[12] Paul White has similarly pointed to a process whereby practitioners underwent a "reversion" in the laboratory, wherein "bestial instincts were unleashed through the repeated and prolonged infliction of pain on helpless creatures." This destabilized the "boundaries between the animal and the human" in the name of clarifying them. Physiologists represented a "divided self," "struggling [. . .] to overcome instinctual sympathies for other creatures in order to fulfill commitments to a higher good."[13] I am going to suggest that we leave such literary fantasies behind as ultimately unhelpful analyses, but that we retain and explore White's last comment about fulfilling "commitments to a higher good," and to determine the ways in which the laboratory space enabled physiologists to practice a historically specific but sharply defined type of sympathy.

I will present the physiological laboratory through the vision, planning, words, and practices of physiologists themselves, who saw in the modern laboratory a theater of "sympathy" that rendered the space also a theater of emotional control. This leads me to the affective practices of scientists in physiological laboratories during physiology's initial rise as a specialist discipline of research. I will explore the common assertion that the laboratory space itself demanded the suspension of sentimental and aesthetic responses to the immediate appearance of suffering, but not in such a way as to permanently affect the emotional sensibilities of the practitioner. The equanimity of the physiologist in his (the gendered pronoun being important) specialist practices of exposing the inner workings of the living being was said to be required in order to bring about new knowledge that would lead to great medical advances. Osler's "nervelessness," that "callousness which thinks only of the good to be effected," or Spencer's vision of more highly developed emotions, were to be ensured by the laboratory itself, its rules, its ethos, its specialized detachment, its equipment, and its special arrangement of space.[14]

The modern archetype of the purpose-built physiological laboratory was envisioned by Carl Ludwig in Leipzig in the 1860s (fig. 4.1). Ludwig founded the physiological institute there, becoming the teacher and mentor of many of the world's leading physiologists, including Edward Schäfer, Arthur Gamgee, William Rutherford, and William Stirling, who would all come to practice in British laboratories.[15] Ludwig was reputed to take precautions in minimizing animal suffering via anesthetics, and maintained that vivisectionists were "the true friends of both man and animal," because they worked ultimately for medical advances that would aid both. They were "well equipped to value the suffering of the vivisected animal." Ludwig would come to lament the antivivisection agitation in England, which he blamed on the English clergy's

Fig. 4.1: Carl Ludwig's Physiological Laboratory, Leipzig (Wellcome Library, London)

meddling in things it did not comprehend, while the institutions of English medicine remained detached from the state. The English clergy, Ludwig stated, "imagines that physiology can alienate the soul," but he determined to prove the "eminent degree of humanity" and the "scientific benefits" of vivisection, all the while remaining the head of the Leipzig animal welfare association.[16]

His laboratory was constructed between 1867 and 1869 to his precise specifications, and was visited in 1870 by the Harvard physiologist H. P. Bowditch. Bowditch's description of the laboratory for the *Boston Medical and Surgical Journal* found its way a few months later into *Nature* (fig. 4.2).[17] With the world's physiological tyros making their way through Leipzig to finish their educations, and with such prominence in the most eminent journals on both sides of the Atlantic, Ludwig's vision became the yardstick for measuring physiological facilities.

The plan of Ludwig's laboratory proceeded as follows, according to the diagram Bowditch included in his report. The building was approximately 199 by 122 feet. Room A was for beginners in microscopy. Room B was the private study of the assistant in microscopy. Room C was for advanced studies in microscopy. Room D was a small library. Room E contained glass cases for storage of physiological apparatus. Rooms F, G, and H were for experimental physiology and contained operating tables, bellows for artificial respiration,

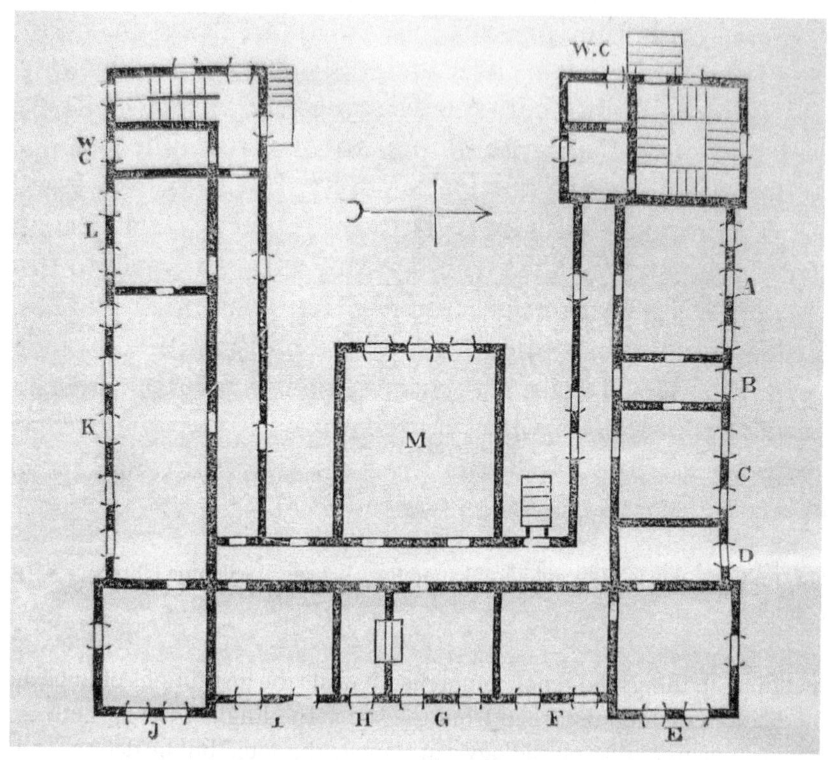

Fig. 4.2: "The Physiological Laboratory at Leipzig," *Nature* 3 (1870)

machines for blood pressure, water baths for controlling temperature, injecting apparatus, evaporating closets, and glass cases. Room I contained two mercury pumps for experiments involving large amounts of mercury. Room J was divided between a weighing room and a room for experiments in acoustics. Rooms K and L were chemical laboratories, and room M was a lecture room with capacity for one hundred students. It was equipped with tables on rails in front of the students, so that the professor could demonstrate his experiments by sliding them along. All of this was on one floor. Below, in the basement, a gas engine powered the respiration and measuring instruments. Here the animals were also stored, with a whole room dedicated to frogs. There was a chamber of refrigerators for chemical experiments, as well as a chamber of furnaces, a workshop, and store rooms. A mechanic lived on site. The courtyard contained arrangements for experiments on horses or other large animals, an aviary, and a fish pond.

With each room's function so clearly defined, and with the division of space so regimented, the physiological student's purpose was defined by the room itself, its specific equipment, its specific scientific end. According to Gerald Geison, Michael Foster's attempt to bring the physiological revolution to Cambridge in the 1870s followed Ludwig's ideal almost exactly. Foster was certain that the design of the laboratory was essential to the correct formation of the student's mind. Physiological instruction had to be practical, hands on, or else the student would never acquire "sound judgement." Only through carefully organized practical experience would a student come to a "just sense of what is right and what is wrong." The best practices of a physiologist—techniques, skills, logical sequences, the handling of apparatus and animals—were to be affective practices, performed with an implicit *feeling* of their rightness, gained through experience. Considering many "of even the most fundamental observations are tedious," the investment in time, labor, and patience, in the right setting, would ensure the right result. The desired result was not only the correct outcome of a physiological experiment, but also the correctly formed physiologist himself. Foster, blighted by the inadequacy of his rooms in Cambridge and bidding for funds for a laboratory along the lines of Ludwig's scheme, noted that he had "students anxious to learn properly," for whom a properly designed space was the key.

Those studying microscopy were to have a table of their own, of which one hundred were to be provided. Foster foresaw a continuous bench, placed in adequate natural light, some three hundred feet long but only twelve feet wide—a "mere corridor." For more advanced students conducting original research, larger tables in a number of small rooms were to be provided so as to permit "quiet concentration." The distinction was mirrored in Ludwig's rooms A and C. For physiological chemistry, Foster desired a laboratory up to forty feet square, with twenty-five benches, with smaller spaces for weighing and spectroscopy, again along the lines of Ludwig's plan. Room for advanced work would be allocated separately. The division of space desired for experimental physiology was similar, made on the basis that it "is impossible for any one to work earnestly at a careful observation in a room occupied at the same time by elementary students."[18] A theme emerges in Foster's vision of physiology: earnestness, tedium, peace, sound judgment. The laboratory was to convey the antithesis of horror.

We find the same sentiments in the speech of George Pollock, upon the opening of the new physiological laboratory at St. George's Hospital in 1887. The modern advantages of medicine—the laboratory space itself and all its obligatory scopes and meters—gave the student "a satisfaction and a confidence," making matters "more plain and smooth," compared with the previous path, which had been "ill-lighted, rugged, and rough." The tedium of physiological measurement was, through the facility of the modern laboratory, "truly throughout a source of pleasure, backed by the conscientious feeling that only one result can be the consequence, and that that consequence is the good of mankind!"[19] Sympathy, benefaction, humanity: these noble affections were channeled to the ends of physiological research through clinical tedium, acutely specialized activities, and acquired judgment. There was not so much a question as to whether a student in physiology had to overcome aesthetic disgust, or a sentimental squeamishness. According to the leading lights who designed physiological spaces, the space itself eliminated any such possibility.

The smoothness of physiological processes was enhanced by the careful incorporation of experimental animals into instrumental mechanisms. With the will to measure physiological outputs—from blood pressure to temperature, from muscle force to nervous energy, matched increasingly by the means to make such measurements—the experimental object increasingly looked like the battery in a machine. Physiological manuals emphasized the output of physiological apparatus—the scratches of a needle in wax—rather than on the input—the excited organs of the living animal.[20] Manuals were designed precisely to describe what the student of physiology would experience in the laboratory. They provided a mitigated way of seeing in advance of the actual practice.

In the depiction of Victor Horsley's laboratory at University College Hospital, for example (fig. 4.3), the highly organized space centrally depicts an experimental animal locked into a physiological apparatus. The close-up segment labels the "Stereotaxic instrument," which permitted precise measurement of the brain, not the animal itself.

This instrumental focus characterized the most famous, or perhaps infamous, physiological manual, *The Handbook for the Physiological Laboratory*. The second volume of this work consisted entirely of plates depicting physiological apparatus, the appearance of histological microscope slides, and experimental animals mid-operation. In general, the animals in these depictions were either shown as instrumental parts of measuring machines, such as the frog in a myograph (fig. 4.4), or as small close-ups of internal parts, devoid of context, color, or character, as in the case of the rabbit's neck (fig. 4.5).

The *Handbook*'s diagrammatical gaze, in common with most handbooks that followed, into the bodies of the frog, the rabbit, and the dog was imagined in such a way as to avoid the aesthetic sensibilities associated with the bloody wound. Rather, furry-edged incisions were simply windows, abstracted from the animal body as a whole, displaying veins, arteries, nerves, ganglions, and glands.[21] The affective practices of the laboratory were prefigured by the mechanical operation manual.

The physiological laboratory was not only a space for practical experience. Students of physiology were also taught basic techniques by direct obser-

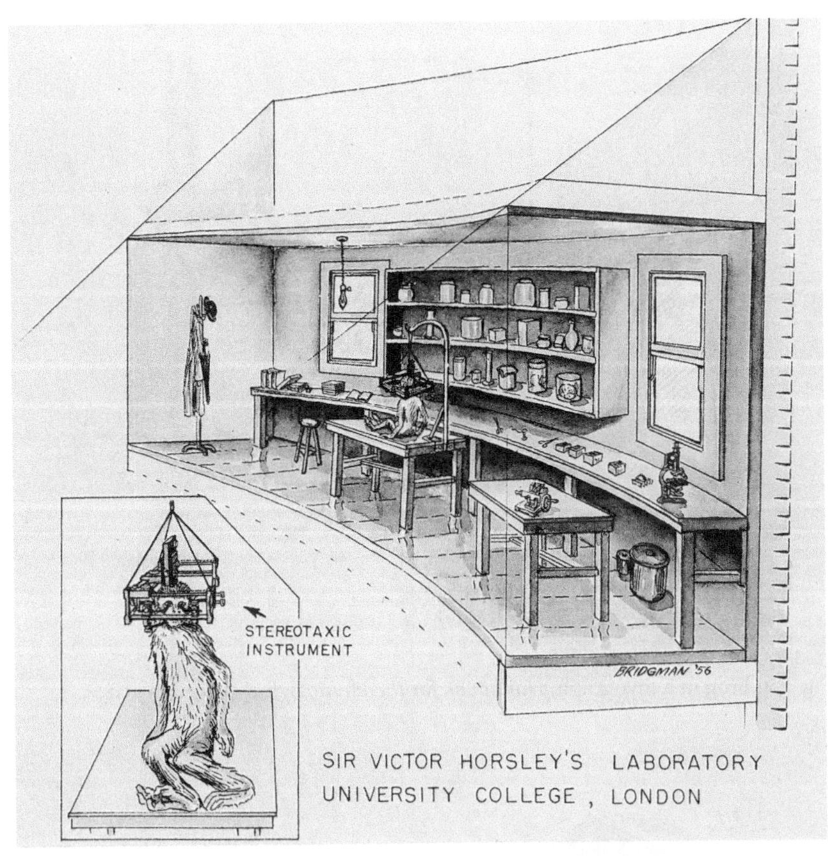

Fig. 4.3: Victor Horsley's laboratory at University College Hospital (Wellcome Library, London)

vation, the importance of which was frequently stressed by professors, as opposed to what might be learned from the pages of a book. Manuals were to assist in practice only, and were, if only read, considered to be useless. Physiologists were unanimous in expressing that only experienced students could be let loose in the laboratory, and this experience in the first instance had to be gained in the direct observation of a master of physiology.

The lecture theater was reimagined for this purpose, to prime the student for the seriousness of the work, but also to ensure the direction of ocular focus. Theaters, when designed specifically for physiology, tended to be high-banked, semicircular amphitheaters, with the stage set for the physiologist and his apparatus. Again, Leipzig was the source of innovation here. Johann Czermak's physiological "Spectatorium" was specifically designed to ensure not only that every student *could* see, but rather that every student *must* see. Completed

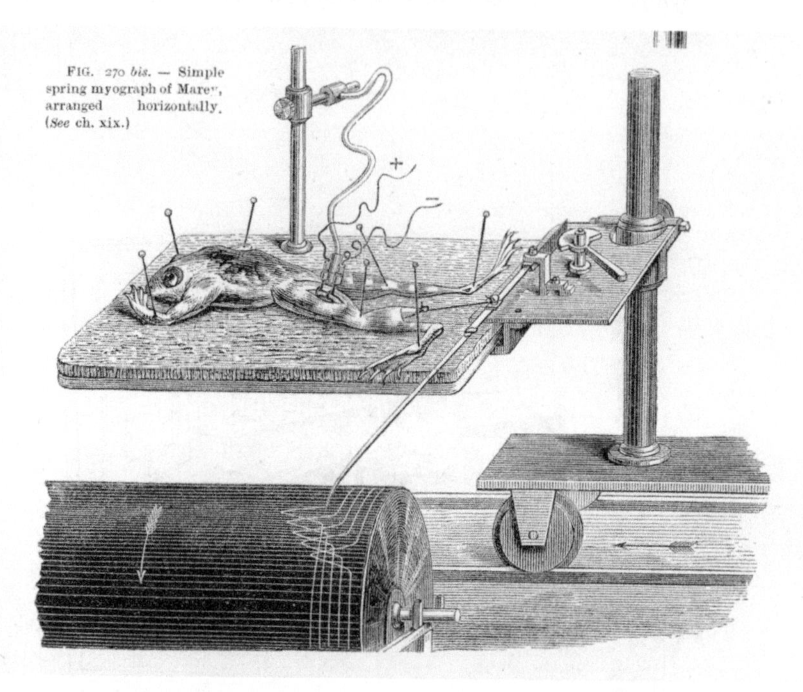

Fig. 4.4: Frog in a myograph, *Handbook for the Physiological Laboratory,* 1873

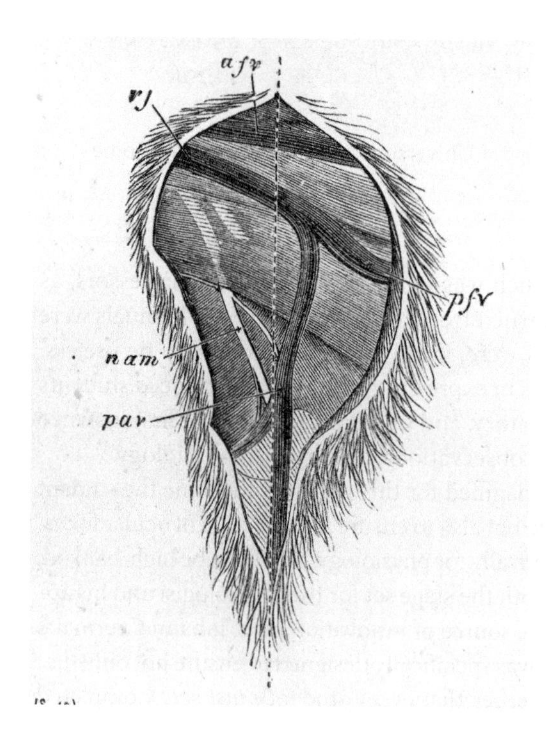

Fig. 4.5: Parts exposed in a rabbit, *Handbook for the Physiological Laboratory,* 1873

in 1873, the amphitheater met the need to present physiological processes to the "direct observation" of the student, with all the conveniences required by the experimenter to hand on the stage, ensuring "educational success."[22] The design, influenced by Czermak's London travels, and based upon anatomy theaters, became archetypical. In England, this kind of space in particular was cause for complaint. The repetition of experiments for teaching purposes was thought to add needlessly both to the sufferings of animals and to the process of calcification of the finer sensibilities of the students who witnessed them. When John Burdon-Sanderson secured £10,000 for the construction of the Oxford physiological laboratory, it was on the understanding that no vivisection would take place for teaching purposes, the practice being limited to original research and his own private practice. In a heated debate that rumbled on through the early 1880s, the Bishop of Oxford illuminated the key point for those who, outside physiology, remained suspicious of it. He spoke

> not only on behalf of suffering dumb creatures, but on behalf of [. . .] human nature, and of its prospects. He did not believe it possible to separate one part of their human feeling from another. He did not believe that a man who had calmly, even joyfully, witnessed the suffering of a victim of vivisection for an hour in a laboratory, could go out into the street and feel a tender compassion for the poor over-laden beast who was being goaded till he fell under his load and died. He did not believe they could be callous to suffering in one form, and deeply sympathetic with it in another.[23]

It was physiologists' contention that precisely this kind of emotional "navigation" was possible.[24] One form of suffering—if indeed an anesthetized animal could be said to suffer—was for medical progress; the other served no purpose whatever and was rightly despised. Such was Sir Henry Acland's response to the Bishop of Oxford, who pointed out that while Burdon-Sanderson was a "great scientific man [. . .] he was more remarkable still for the quality of gentleness." Such a man would "rear up for them in this place [. . .] by his skill and tender nature, men who should be an example of how these things should be learnt, and who should go forth from this University as Christian men."[25] The building was approved, with an annual £500 grant for working expenses, and was completed by 1885.[26]

The customs, procedures, apparatus, and telos of the physiologist seemed to be inaccessible to the outside observer's gaze. The Old Brown Dog affair that began in 1903, of which much has been written, began life as a set of competing narratives about what happened in a physiological lecture conducted by William Bayliss and Ernest Starling (fig. 4.6).[27]

Significant to the anti-vivisectionists' case was the disposition of the assembled scientists, who were reported to be in a state of levity, "a spirit of

Fig. 4.6: The Old Brown Dog affair, 1903 (Wellcome Library, London)

jocularity."[28] A similar accusation had been leveled at David Ferrier in 1875 on the occasion of a lecture on "Experiments on the Brain of Monkeys," at which the audience laughed throughout at the descriptions of the monkeys' grotesque movements and facial contortions. The secretary of the Royal Society for the Prevention of Cruelty to Animals, who was present, "left the room in consequence of the pain with which he saw the laughter of the young people."[29] On both occasions, the moral economy of the physiological lecture was breached because of a peculiarly British tendency to engage public interest in, or even assign jurisdiction over, the scientific world. What constituted humor in the physiological laboratory—in accord with Pollock's notion that physiological work could be a "source of pleasure"—could not easily be translated or explained to the layman; nor could there be any reasonable expectation that the layman's emotional reactions to physiological space and its associated practices could be controlled. So long as science was held up to public opinion, and a certain lens of public scrutiny, there would be this incommensurability of moral economy, a divide deepened by the increasing refinement of physiological specialism.

In 1988 Steven Shapin set out systematically to study the "venues of knowledge," or the sites of "knowledge production," hoping to discover how "transactions" across the thresholds of such spaces were managed. His focus was scientific experimentation in the private realm of gentlemanly domesticity in the seventeenth century, arguing that the "word of a gentleman, the conventions regulating access to a gentleman's house, and the social relations within in it" were what "underwrote assent to knowledge claims." This, so the argument goes, is in stark contrast to "more modern patterns," where scientific credibility has to do with the "visible display of the emblems of recognized expertise," vouchsafed by other experts.[30] In general, the shift from the gentleman scientist, based at home, to the professional scientist, who went to *work*, in a *specialized* place, is thought to have taken place from the mid-nineteenth century. Physiology in particular usually features prominently in this story, as a prime mover in the transition to specialized professionalism in science.[31]

Yet with any transition we should expect the edges of two distinct realities to be blurred. I have shown elsewhere that the importance of the gentleman scientist overlapped significantly with the era of the paid scientific employee, especially in England.[32] It should not be surprising therefore that the gentleman's various spaces of activity should also have been brought into the era of professional laboratories. Both Ludwig's and Czermak's revolutionary realizations of a new professional space for physiology came with living quarters. Ludwig constructed his domicile above the purpose-built laboratory, with space not only for his family, but also for the "other persons connected with the laboratory."[33] While not all physiologists could boast of such a space, doubtless the lines between private and professional space were still frequently blurred. In Emanuel Klein's testimony before the Royal Commission on Vivisection in 1876, infamous for its claims to complete indifference to animal suffering, Klein acknowledged that he had conducted physiological experiments at the Brown Institution in his "private room," adding: "Those that I do for teaching purposes, physiological purposes, I do in my private room; I live there."[34]

Even if the home, strictly considered, was no longer the center of experimental life, it may have been the center of professional discourse. The Physiological Society, for example, which brought together the leading lights of the profession in order to defend it, was established at John Burdon-Sanderson's house on Queen Anne Street in London, and thereafter meetings rotated among the homes of the various scientist members, crowding parlors and no doubt adding an extra burden to the domestic economy.[35] And laboratories themselves were softened by domestic touches—the hat and coat stand in Victor Horsley's laboratory, or the laboratory pet dog seen wandering among the rabbit cages in Pasteur's physiological classroom (fig. 4.7).

The proximity of physiology to domesticity, whether practically, materially, or discursively, almost certainly made the navigation between emotional

Fig. 4.7: The Physiological Classroom of Louis Pasteur (Wellcome Library, London)

and moral worlds more straightforward. Entrance to the laboratory signaled a change in emotional gears, where a conscious application of emotional control was enacted through physiological practices and embodied in instruments, apparatus, and animal objects, all of which were endowed with the telos of an ultimate good. What made that good tangible was the proximity of the physiological world to that of loved ones—family members, friends, and indeed, pets. So many physiologists, daily performing vivisections, declared themselves to be animal lovers. We make a historical error if we adjudge them monstrous hypocrites. On the contrary, the configuration of physiological space as part of a moral economy of physiology enabled vivisection to be, for them, a precise, important, and beneficent expression of that love. *Fin-de-siècle* physiology, attempting to secure civilization's salvation, dwelled firmly within civilization. Its specialized spaces, the laboratory and the lecture room, facilitated and even required the equanimity of practitioners. But it was nevertheless intimately connected with a world beyond, to the moral economy of the home, to a Darwinian ideal of humanity and progress, and to the idea of social good.

Aesthetics and Anesthetics

What is the appropriate emotional response to the sight of the surgically opened, living body? As we have seen in chapter 3, for the opponents of Darwinian morality, it was essential that the sight of blood and gore immediately aroused disgust, which might be expressed, from a distance, as pity or compassion. The spaces of physiology were themselves designed to make possible the *routine* of vivisection, both the methodological procedures involved and the affective practices of the physiologist, carrying out these procedures as "ordinary," run-of-the-mill activities. But much of the argument against vivisection centered on those procedures themselves, and what feelings they ought to have brought to the surface. What were physiologists' own reflections on their emotional conditioning, preparing for and working through the aesthetics of the opened body? How did knowledge and application of anesthetics affect this emotional conditioning? Here I want to analyze scientific coping strategies, through which the aesthetics of pain, that is, the *sight* of suffering (whether or not any actual suffering took place at all), was mitigated, justified, rationalized, and subjected to emotional control. I will argue that a diminution of the aesthetic response to the sight of blood, in conjunction with knowledge of anesthesia, allowed physiologists to conform to their new moral code, abstracting sympathy to make it a response to suffering in general, removed from the immediacy of the laboratory, and in the name of "humanity." Whereas the surgeon did this for the singular good of an individual patient, the physiologist appealed to a higher morality: His operations were for the good of everybody.

Paul White has suggested, with regard to vivisection, that the "crux of the late-Victorian debates was not just whether particular feelings were present in the experimenter or the animal, but the nature of emotion itself; its role in science and medicine—and in human society generally—seemed open to question."[36] Testing the historiographical credence given to the hardened heart of the late-Victorian scientist requires an investigation into what physiologists thought about causing (or avoiding causing) pain in animals.[37] Where Patrizia Guarnieri talked of the Jekyll and Hyde character of the scientist, it is noteworthy that she described the "white-collared scientist" monster holding down "an etherised dog." It is surely necessary to ask whether the use of anesthetics made any difference to the emotional experience, the affective practice, of holding down (actually, an etherized dog would not need any restraint) an experimental animal. I am operating on the assumption here that physiologists exercised logical consistency in their ethics and practices. It is important to understand how someone might have claimed to have been a lover of animals, to detest cruelty, and be a practicing physiologist at the same time.

To what extent were experimental animals thought to feel pain? Where did that pain weigh in the balance of comparative suffering? The answers to these questions allowed medical scientists to rationalize their own feelings in response to the experience of (inflicting) animal pain. G. M. Humphrey, professor of anatomy at the University of Cambridge, told the Royal Commission that the comparative smallness of animal nervous systems indicated that they could not possibly suffer so acutely as humans. Moreover, signs of a struggle were not construed as reliable indicators of pain. The "violent contortions of the worm" on a hook did not necessarily indicate pain, "for there may be violent contortions and no suffering whatever." So much, Humphrey said, had been learned from the painless muscular excitations of men under chloroform, which looked like pain but were not, as well as from the painless convulsions of epileptics.[38]

This commonly stated opinion captured physiologists' distrust of the outward signs of pain, which might otherwise have led to unwanted or inappropriate emotional responses to it.[39] Such responses were deemed part of a culture of sentimentalism against which physiology aligned itself. Physiologists believed that the lack of pain in the animal removed any objections on the grounds of taste, and they saw the emotional pain of anti-vivisectionists under such conditions as nothing more than a sentimental (feminine) reaction. James Crichton-Browne had defended Ferrier, whose lectures on his monkey-brain experiments had proven so controversial, and with whom he worked at the West Riding Asylum, in precisely these terms. The outward signs of pain could be achieved in animals without a brain, "or in the deepest state of anaesthesia," by a simple "stimulation of the motor centre." The apparent "intense and protracted agony" was "not greater than that of a pianoforte when its keys are struck."[40] Emotional control was achieved by the physiologist by applying what he knew as a filter against the bubbling up of instinctive or aesthetic sympathetic reactions.

According to George Burrows, who was president of the Royal College of Physicians, only a "very limited number of experiments [. . .] will cause a degree of pain to the animal," and under those circumstances it would be "painful to the operator and to everybody else to contemplate."[41] Sympathy in the immediate setting of the laboratory was therefore rationally limited. James Paget, trusting in the "general humanity of scientific men," thought such men could be "left to be fair judges of what amount of pain it is reasonable to inflict for the sake of attaining some useful knowledge."[42] William Sharpey, for example, was convinced that experimentation did not have "the effect of blunting the feelings" or "hardening the nature" of physiologists.[43] The common concern that vivisection tended to brutalize the operator could be dismissed on the basis that animals' exposure to pain was minimized, for some of them by their own lowly nervous systems, and for others by the use of anesthetics.

Even after the use of anesthetics was prevalent, comparative capacities of sensitivity to pain were continually used to justify experimentation, perhaps because anesthesia was not deemed appropriate for every experiment.[44] "The sole means," according to Edmund Gurney, of arriving at a "conscientious estimate of others' suffering [. . .] lie in imagining it as one's own." The anthropomorphism of this cross-species sympathy raised the suspicion that animals were commonly allocated a greater capacity for experiencing pain than their physiologies warranted. Gurney argued for a "close relation of suffering to intelligence."[45] Intellect was the key factor that enhanced suffering, and humans—even to the ardent utilitarian—were thought to have the largest share. Some animals shared the physiological systems of humans, but their brains were "in proportion to the rest of the body, very much smaller than in the case of man."[46] Given the likely benefits derived from physiology, vivisection could thus be justified.

These utilitarians had a good precedent for proceeding in this manner, for J. S. Mill had long-since said that a "being of higher faculties requires more to make him happy, is capable probably of more acute suffering, and certainly accessible to it at more points, than one of an inferior type." It was, after all, worse to be a human being in pain than a pig in pain; worse to be Socrates in pain than a fool in pain.[47] The twist was to say, with one eye on the anti-vivisection movement, that if anybody thought differently about the pig or the fool on behalf of the pig or the fool, they were guilty of a category error, for in fact these advocates only knew their own side of the equation.

At the International Medical Congress (IMC) held in London in 1881, the largest ever assemblage of eminent medical men from around the world to that date, John Simon gave a widely heralded speech defending medical science. He particularly denounced the aesthetic sensibilities of anti-vivisectionists: "In certain circles of society," he said, "aesthetics count for all in all; and an emotion against what they are pleased to call 'vivisection' answers their purpose of the moment as well as any other little emotion." The medical profession could not seriously argue with such people, for they did not share a moral standard, or a world view: "Our own verb of life is *εργαζεσθαι* [to work], not *αισθανεσθαι* [to feel]. We have to think of usefulness to man. And to us, according to our standard of right and wrong, perhaps those lackadaisical aesthetics may seem but a feeble form of sensuality." But that was not to say that he felt nothing with regard to his work. On the contrary, he thought of inflicting pain "with true compunction," but he did it nonetheless because of the "end which it subserves": the promotion of "the cure or prevention of disease in the race to which the animal belongs, or in the animal kingdom generally, or (above all) in the race of man." Under such conditions he would not "flinch" from this "professional duty, though a painful one." Simon was referring to his own pain.[48]

Concerns about causing pain and feeling sympathetic pain should have been put to rest by the widespread use of anesthetics, which were employed in the vast majority of experiments after 1876. The primary benefit of anesthetics was not that the experimental animal no longer suffered, but that the major concerns of the physiologist were alleviated: the greater good could be sought unhindered, the operator would not lose his nerve, and he would safeguard his "feeling" heart. On a practical level, it also meant that the animal would keep still, though this fact was seldom mentioned.[49] Anesthesia objectified the experimental subject, allowing physiologists methodically to remove emotions, not *from* themselves, but *to* more distant, abstract objects. Without anesthetic, the experimental animal's status as a sensitive being could involve it in a reciprocity of aesthesia, of physical pain in the animal and the reflection of that pain—sympathy—in the operator. This might inhibit the researcher in beginning, or in pursuing the ultimate ends of his research. As Carolyn Burdett has recently argued: "Aesthetic response belongs in the relation between viewer and object, as a consequence of what the object precipitates or excites in the body of the viewer. What the viewer then experiences (the consequent feelings or emotions), they then project back and experience anew, as if located in the object."[50]

Indeed, not to feel this sympathetic pain might be a sign of brutality, giving rise to the "general accusation of hardness" to which medical science was accustomed.[51] Chloroform and ether were safe ways to cut this reciprocal aesthesia, replacing it with a similar but opposite reciprocity of *an*esthesia that could preserve both the nerve and the tenderness of the operator.[52] The benumbed object excited nothing in the viewer (operator), eliminating the possibility of projecting sensation back into the object. As such, William Carpenter averred that "removing" pain had become a "matter of duty" for physiologists, who could project their sympathetic gaze outside the laboratory.[53] By rendering the experimental subject as object, emotions were removed from the physiological procedure, in the name of a more abstract "humanity."[54]

There is a wealth of evidence to demonstrate that physiologists knew they were doing exactly this, even though they may have thought it possible without anesthetics.[55] John Burdon-Sanderson, coauthor and editor of the *Handbook for the Physiological Laboratory*, averred his belief in a certain capacity inherent in the highly evolved civilized male. A man, much more so than a woman, was capable of "directing mental effort to a recognized purpose" without succumbing to the "greatest enemies," those "emotional or sentimental states," including sympathy, which so often "handicapped" women in their endeavors. A scientific man was singularly well-equipped for a "life directed to the fulfilment of a recognized purpose to which others must yield."[56] Burdon-Sanderson famously neglected the subject of anesthetics in the *Handbook*, and was repeatedly asked to justify the infliction of pain in the physiological laboratory, which he did by reference to "the circumstance that we are working for an important and good object."[57] But if the infliction of pain could be justified if there was "a certainty that the human race would be benefited by it," how much more easily could an experiment be justified under anesthesia?[58] Burdon-Sanderson acknowledged that he "should condemn the nonemployment of anaesthesia" wherever anesthesia could be used, and indeed acknowledged that he had failed in not making this clear in the *Handbook*.[59] Yet he remained convinced that responsibility for ensuring the "greatest possible result," "at the expense of as little suffering as possible," lay with the scientist himself.[60] It might even be argued that the failure of the *Handbook*'s authors to make humanitarian overtures toward those whom Burdon-Sanderson would have adjudged to have succumbed to their "emotional or sentimental states" was consistent with an imperturbable direction of mental effort.

Another of the *Handbook*'s authors, the noted Scottish physician Thomas Lauder Brunton, also expatiated on the special qualities of the scientist, making the distinction between two types of compassion. Both medical scientists and anti-vivisectionists were "anxious to lessen the amount of pain and suffering in the world," but where one looked to "the immediate and designed suffering of a few score of animals," the other looked to "the ultimate relief of the undesigned pains of disease in animals and in men." To civilized people, Lauder Brunton admitted, the "mere sight of suffering is painful." This "painful impression" causes some immediately to turn away and thus "be rid of the disagreeable feeling." For others, "it excites a desire to relieve the pain of the sufferer, however disagreeable, disgusting, or trying the task may be." He put physiologists in the latter group. Such a "power of controlling one's own emotions, of disregarding one's own feelings at the sight of suffering," varied from person to person, but it could be trained. It involved subordinating emotion to judgment, and it was aided in the case of physiology by practice, knowledge and anesthetics. The daily experience of experiment would, in itself, help with the process of putting judgment before feeling, allowing these "humane men" to "purchase future good at the expense of present pain." But since the "great majority" of experiments were "rendered painless by means of anaesthetic agents," physiologists could, with measured judgment, learn "to disregard their own feelings, and to concentrate their attention on the interests of the [human] patient."[61]

This had been forcefully asserted by the institution of medicine at large as early as 1881. The IMC in London unanimously passed a resolution that had been drawn up under the auspices of the Physiological Society. It recorded that body's "conviction that experiments on living animals have proved of the utmost service to medicine in the past, and are indispensable to its future progress." It strongly deprecated the infliction of "unnecessary pain," but demanded "in the interest of man and of animals" that "competent persons" should not be restricted in their experiments.[62] In addition, many of the age's most prominent medical scientists and physicians came forth with their own similar defenses. Gerald Yeo, professor of physiology at King's College London, underscored the profession's abhorrence at the infliction of pain by laying before the public an extended analysis of the prevalence of anesthetic usage, setting out to prove that there was no "want of tenderness amongst English physiologists" and that "Pain forms [. . .] but a rare incident in the work of a practical physiologist." William Gull emphasized the "moral duty" of investigating "problems of the highest importance to mankind" when the "solution of these problems is within the scope of the human intellect." This course by no means made physiologists "indifferent to or careless of inflicting pain." Their character had already been safeguarded by the 1871 resolutions of the British Association, the first of which read: "No experiment which can be performed under the influence of an anaesthetic ought to be done without it." It was with happiness that Gull noted that the "great majority" of experiments on the nervous system "are performed on decapitated frogs, or on other animals under the influence of anaesthetics."[63]

Physiologists, as a body, were pain-aware, mindful of the freedom given to them by anesthetics and focused on what they perceived to be the higher moral ends of their operations. Those moral ends, understood as the alleviation of suffering on a human scale, were embedded within the moral theories of Darwin and his contemporaries. Yet there was still emotion work to do, even for the most ardent Darwinist. The emotional and moral compasses of the theorist were not automatically calibrated to make vivisection work possible, or palatable.

Evolutionists and Physiology

The link between physiology and evolutionary ethics is abundantly clear, and Darwin himself worked behind the scenes in collaboration with John Burdon-Sanderson, John Simon, T. H. Huxley, George John Romanes, and others to ensure protective legislation for physiologists.[64] Huxley served as the most notable defender of vivisection on the committee of the Royal Commission on Vivisection, while elsewhere publicly denouncing "the venomous sentim[ent]ality & inhuman tenderness of the members of the Society for the infliction of cruelty on Man—who are ready to let disease torture hecatombs of men as long as poodles are happy." Herbert Spencer is reputed to have regarded vivisection to have been "so justified by utility to be legitimate, expedient, and right," on the condition of state supervision.[65]

Robert J. Richards has clearly demonstrated that Darwin's evolutionary ethics was "a morality of intentions." This meant judging moral action not on what was done, in abstraction, but on the intended outcome. To better do this, according to Darwin, "we must look *far forward* & to the *general action*—certainly because it is the result of what has *generally* been best for our good *far back*."[66] The loose body of evolutionary scientists characterized anti-vivisectionists as adherents to a "false" or "mistaken" humanity because they allowed their conduct to be led by an immediate reaction to what they saw, or sensed, as wrong, without due consideration for what was actually good for humanity. The application of Darwin's own moral theory to the matter of vivisection is startlingly clear. In his most famous contribution on the subject, Darwin wrote of the "incalculable benefits which will hereafter be derived from physiology, not only by man, but by the lower animals [. . .] In the future every one will be astonished at the ingratitude shown, at least in England, to these benefactors of mankind."[67] For Darwin, anesthetics were morally desirable, but once used there could be no remaining objection to vivisection, a term he wished to replace with "anaes-section" to clear up any moral doubts.[68] Even without anesthetics, an operation could be justified "by an increase in our knowledge," and could give the operator protection against the "remorse" that would otherwise arise from his procedures.[69] The evolution of sympathy allowed the "surgeon to harden himself whilst performing an operation, for he knows that he is acting for the good of his patient."[70] With such Darwinian license, one might think a devotee could take up the scalpel without compunction. The reality was never so straightforward. For some it caused an agony of despair.

What follows is a small case study of the creation of a scientific self, that is, a person's understanding of himself as a right thinking, right feeling, morally acting *practitioner* of science. It concerns the process by which scientific activities—experimenting, teaching, writing, collaborating—are imbued with moral and affective—that is to say, *virtuous*—qualities. To *cut* here is to do the *right* thing, for the knowledge gained by it will result, somewhere down the road, in a lessening of human and/or animal suffering; ergo it is an act of sympathy; ergo I *feel* sympathetic. And conversely, but simultaneously, I *feel* sympathetic, according to a certain theoretical prescription, for the mass of human and animal sufferers in the abstract; ergo I express this sympathy by *cutting* here, for this is the only *right*, the only *moral*, course of action. Such a process is fraught with difficulty, since it takes place in the context of habitual emotive processes that run counter to it. To cut an animal is *wrong*, an act of *callousness*, and causes an *increase in suffering*. These prevailing prescriptions, polar opposites of the new science of sympathy, were highly successful. A scientist embracing the new principles had to address his own inner responses to the thought or the act of vivisection. The emotive process prevailing in the wider moral economy of society matched a visceral feeling of disgust to an expression of outrage. It is, arguably, still the dominant emotive when dealing with vivisection in contemporary society. To control, negate, or dismiss those feelings of disgust involved a new emotive process. To highlight the "emotional crisis" involved in this kind of self-formation, against a chorus of societal disapproval and misapprehension, I will focus on two eminent scientists, one with a deep-seated fear of wielding the scalpel on the living, and the other with a real difficulty in really feeling Darwinian.

When Thomas Henry Huxley addressed the British Association as president in 1870, he could not have foreseen the extent to which he would become personally involved in the campaign to defend vivisection over the next two decades, against an increasingly vociferous activism that sought to tar physiologists with the brush of cruelty, brutality, and heartlessness. Huxley's engagement with the vivisection question affords a perfect example of the formation of a scientific self, a process involving a careful negotiation between his inner emotional reaction to the idea of vivisection, marked by disgust, and his considered emotional expression of support for vivisection, based on a learned notion of sympathy. Huxley's public and private record of his conflicted feelings about vivisection afford us a perfect example of an emotive process at work.

Thomas Henry Huxley was a champion of Darwin, experimental science and medicine, and education. When Huxley went on the attack he did so fiercely. When defending a principle he was resolute. As president of the British Association in 1870 Huxley went on the offensive against those who cried foul against vivisection, taking advantage of the presence of Dr. Brown Séquard, whose own speech had depended on the results of vivisection research. In his speech before the association, Huxley helped form what would become a well-rehearsed line of argument for physiologists over the following two decades. The British Association had, not long previously, faced a resolution aiming to pledge the association "to abstain from making grants of money to persons engaged in experiments which involved vivisection." Holding up the example of Brown Séquard, he asked his audience if this was a man "likely to inflict one particle of pain upon any creature whatever without having a plain and definite purpose in view." Huxley asserted his own aversion to the thought of "pain in anything." Together, they were far from "indifferent to pain," and "in no sense were they cruel." The suggestion that vivisection should receive no institutional support must, he thought, have arisen out of ignorance; for the diagnostic knowledge acquired by Brown Séquard through experimental physiology had since been applied to medical practice, and "before long his wonderful mastery over symptoms caused his consulting rooms to be absolutely crowded by human beings suffering under multiform varieties of nervous disorders, who sought at his hands and from his knowledge that which they could not obtain elsewhere." Not impugning the cause of preventing animal cruelty, Huxley nevertheless averred that to conflate the "brutal violence of the carter or the wife-beater with an experiment carried out by a man of science, gently and for the purpose of relieving misery, the enthusiasts in that cause should change their name, and convert themselves into a society for the promotion of cruelty to mankind." This quality of gentleness, for the sake of diminishing suffering, was the key. The object of vivisection, never mind its means, was sympathy on a grand scale. Those means could well be overlooked when considering that "certain kinds of truth were only attainable" by such experiments, with the result that "the welfare of thousands and thousands of untold human beings who might otherwise be suffering unimaginable misery" would be improved.[71] Physiology, for Huxley, was a practice of sympathetic kindness based on a deep scientific knowledge of the causes, and medical remedies, of suffering in humanity.

The connection between animal physiology and human physiology that made vivisection useful depended on Darwin's work on the evolution of species from a common ancestor. To subscribe to evolutionary theory—and there was no more vocal subscriber than Huxley—almost automatically implied support for vivisection, since the benefit of the practice for human health seemed self-evident. Huxley affirmed this to Darwin personally, agreeing with Darwin that vivisection was "a matter of right and justice." Of course Huxley, as with Darwin, absolutely demanded the use of anesthetics wherever possible, and publicly and privately sought to eliminate "unnecessary" suffering in order to take "the wind out of the enemy's sails."[72] But the central equation tended to be held in common among physiologists, biologists, and theoretical scientists: a small amount of suffering ought to be permitted if it eliminates a larger amount of suffering by its results.

The problem for Huxley was that, although he agreed with this central tenet of Darwinian sympathy, he himself did not have the stomach for it. He defended physiologists out of a sense of "duty" because they added "immensely [. . .] to the means of alleviating human suffering, against the often ignorant and sometimes malicious clamour which has been raised against them." Yet he had to confess that "personally, indeed I may say constitutionally, the performance of experiments upon living and conscious animals is extremely disagreeable to me."[73] According to his son, even if Huxley "did not care to undertake such experiments personally, he held it false sentiment to blame others who did disagreeable work for the good of humanity."[74] When faced with Klein's testimony before the Royal Commission in 1876—Huxley was the most active supporter of vivisection on the committee—he threw his hands up in despair, writing to Darwin that, "I did not believe the man lived who was such an unmitigated cynical brute as to profess and act upon such principles, and I would willingly agree to any law which would send him to the treadmill."[75] The situation was desperate, and this prompted Huxley to invite Darwin himself to testify to vivisection's utility before the commission, which he did.

Huxley's inner turmoil over vivisection was heightened by his acute awareness of the importance of physiological demonstration for the purposes of teaching, even his own. He ran into trouble in 1874 over the content of his physiology summer course at South Kensington. His lectures were approved, but vivisection was banned, citing concern about the demonstrations Huxley had carried out two years previously in his Courses to Teachers.[76] Huxley had already been "anxious" about the prospect of physiological demonstrations, but knew that without them "the subject could not be properly taught." When Huxley had defended Brown Séquard at the British Association, the audience had pilloried him. The memory of the abuse stirred in him as he planned his syllabus, and he fully expected "every description of abuse and misrepresentation" if he went ahead with the demonstrations.[77] But if the likely outrage of society weighed with him momentarily, he ultimately dismissed it as irrelevant. Huxley could well imagine the emotional responses of society to the public display of vivisection, for these emotional responses lay in his own breast. He suppressed them by reference to the abstract reasoning that justified scientific experimentation. He clearly engaged a theoretical emotional prescription for scientific practice and used it to beat down the opposing feelings that were, to a large extent, borne on the tide of the prevailing emotional regime at large. Huxley found strategies practically to negotiate this emotive process, so that his public expression of sympathy with physiologists and the abstract mode of Darwinian sympathy per se did not jar too violently with the feelings of disgust he felt at the prospect of cutting into a living animal.

The principal method employed by Huxley to displace his anxiety was to get someone else to do the practical work. He hired demonstrators to assist in his lectures, so that he did not have to be personally involved in any experiment. He gave them strict instructions about the administration of anesthetics or the prior destruction of the brain (presumably for experiments on frogs) and then left them to it. This was to his satisfaction, for he believed he had taken "such a course as I believe is defensible against everything but misrepresentation."[78] Part of Huxley's attempt to shore up this emotive failure—that is, the lack of concord between inner feeling and outward expression—was to refuse to dwell on his personal misgivings in public. Unfortunately for Huxley, this often had the effect of increasing the public's sense of ire concerning his advocacy on behalf of physiologists.

An example of this concerned Huxley's extremely popular (and lucrative) textbook, *Lessons in Elementary Physiology*, first published in 1866.[79] The book, designed for beginners in the study of human physiology, opens with an account of the necessity of observation to ascertain the requisite knowledge. In what almost seems like a throwaway comment, Huxley points out that such observations would be "easy," "if the bodies of men were as easily procured, examined, and subjected to experiment, as those of animals."[80] This passive acceptance of the utility of animal experimentation was elaborated in the second edition, which pointed out that while "mere reading" could verse a student in "knowledge of science," a "knowledge of a very different kind" could be attained by "direct contact with the fact." Huxley worried that "the worth of the pursuit of science as an intellectual discipline is almost lost by those who seek it only in books." To acquire *practical* knowledge, a student needed to acquaint himself with the "physiological anatomy and histology, the organs and tissues of the commonest domestic animals," which afforded the student "ample materials." Huxley referred to the direct analog of human organs in animal organs—the kidneys, eyes, hearts, and lungs of sheep, for example—and of the similar properties of "living tissues" in the frog.[81] In later editions, this was augmented by illustrations "from the Rabbit, the Sheep, the Dog, and the Frog, in order to aid those, who, in accordance with the Second Edition, attempt to make their knowledge real, by acquiring some practical acquaintance with the facts of Anatomy and Physiology."[82]

During the crisis of 1876, when opponents of vivisection were highly motivated to find evidence of the corrupting influence of physiology on the moral fiber of its practitioners, Huxley's textbook came under the spotlight. The 7th Earl of Shaftesbury took up a report in the *Record* newspaper that had charged Huxley with "advocating vivisections before children, if not by them."[83] In fairness to the *Record,* and to Shaftesbury, it is not too difficult to see how they might have arrived at this impression. Huxley went public, advising Shaftesbury in *The Times* that "physiological anatomy" was "not exactly the same thing as experimental physiology" and that the recommendations in his textbook might be carried out without resorting to a single "vivisection," that is, the cutting of a living animal.[84] Experimenting with pithed frogs was akin to practicing on dead animals. The majority of dissections were on dead organs. Yet one cannot help feel a certain disingenuousness in Huxley's ill-tempered denials. His earlier trouble with his summer courses had rested precisely on the fact that Huxley thought it essential to have physiological demonstrations performed before students, his "dislike to the infliction of pain both as a matter of principle and of feeling" notwithstanding.[85] Some of those experiments certainly involved animals under anesthesia, which Huxley could hardly have explained away as distinct from experimental physiology.

The real difficulty for Huxley, which better explains his indignant response, was the thought that he might credibly be tarred with the label "cruel," "heartless," "callous," or be thought to be generally wanting in feeling. Because he considered himself highly sensible of the suffering of others, including animals, and because such sensibility was a social requirement of a public figure, vivisection threatened to bring down not only the reputation of science but, more importantly for Huxley, the reputation of T. H. Huxley himself. Here the emotive process becomes particularly fine grained, for Huxley was able to surrender neither his tender feelings, in accord with societal expectations and prescriptions of a refined man, nor his attachment to the theoretical and indeed moral underpinnings of the science that had made his name. Caught between the two, Huxley had to try to reconcile his scientific principles and the experimental practices of his peers and friends with a deeply held inner feeling that the infliction of pain, or even the spilling of blood, was abhorrent to him.

For Huxley this meant a staunch appeal to the rational on the one hand and a careful reconceptualization of the connection of emotions to moral character. Huxley wrote in the "Progress of Science" of the risk that "the fanaticism of philozoic sentiment"—a classic Huxleyism for the misplaced emotional zealotry of the anti-vivisectionist cause—might overpower the "voice of humanity."[86] This asserted a clear priority for human welfare that was in perfect accord with even the arch advocates of Utilitarianism.[87] Provided that the love of one's neighbor was not superseded by the "love of dogs and cats," Huxley believed that "the progress of experimental physiology and pathology" would "indubitably, in course of time, place medicine and hygiene upon a rational basis."[88] For those who pointed to the infliction of pain, Huxley pointed to the question of motive or intention. "The wanton infliction of pain on man or beast is a crime," he told a student in 1890, but he reminded his correspondent that "the criminality lies in the wantonness and not in the act of inflicting pain *per se*." In this, Huxley was again in accord with the Utilitarian masters of his own time, but also strictly in accordance with the animal cruelty laws that had been developed in England throughout the nineteenth century. Pain for a fit purpose could be inflicted without damaging the emotional capacity for tender feeling or the moral fiber of the perpetrator. This, even according to Huxley's own appraisal of the moral and emotional difficulties of the question of vivisection, involved subjecting immediate emotional responses concerning physiological practice to abstract reasoning as to the ultimate end of sympathetic action. He knew that "the practice of performing experiments on living animals is not only reconcilable with true humanity, but under certain circumstances is imperatively demanded by it." Huxley himself could not master the emotive sufficiently to carry out vivisections himself, but he understood the importance for physiologists of successfully beating out their own inner feelings in favor of scientifically derived humanistic prescriptions: "[I]t has been my duty to give prolonged and careful attention to this subject, and putting natural sympathy aside, to try to get at the rights and wrongs of the business from a higher point of view, namely, that of humanity, which is often very different from that of emotional sentiment."[89] Indeed, this was emotion on a higher plane. This was instinct taken in hand, guided, reframed, redirected. In this, Huxley exemplifies the formation of scientific self, the appropriation of theory, and its translation into practice and lived experience. From the evolutionary blueprint for the emotional basis of moral conduct, taken from the work of Darwin, Huxley derived a guiding set of principles for how to feel, act, judge, work, and live. He did not always live up to them.

Neither did Romanes, Huxley's polar opposite. I have shown elsewhere that Romanes was not only a chief adjutant in the campaign for physiology and vivisection, but also a pursuer of physiological research in his own right, performing vivisections in a private capacity, and offering up his own dogs as potential subjects.[90] He suggested that Darwin write a pro-vivisection article for the *Nineteenth Century* entitled "Mistaken Humanity of the Agitation: Real Humanity of Vivisection." Various examples throughout this book have shown Romanes to be a standard bearer for the interpretation, dissemination, and practice of morality as constructed by Darwin in his *Descent of Man*. In public matters, Romanes consistently answered the critics of science, defending the humanity, sympathy, and feeling of physiologists in particular. Identifying strongly as a gentleman, it was essential for Romanes to believe and embody these things. Romanes's problem was that he could not easily relinquish "common compassion" in a religious idiom. He began his career with a weighty treatise on the power of Christian prayer, before discovering a hero in Darwin and dramatically altering his world view. His next work was a *Candid Examination of Theism,* in which all his major principles were turned on their heads. It was published anonymously, and might be viewed as a particularly public example of introspective hand wringing, or of personal growth laced with doubt. In his first book, Romanes had condemned positivism and the materialism of the scientist, asking, "Are we to relinquish this great and hallowed creed, merely for the sake of an empty figment of the intellect, which can have no substantially valid reason for its support?"[91] Faith had been his rock. Within four years, Romanes was finding materialist objections to every aspect of theism. His intellect had trumped his heart, and he confessed as much. It caused him the "utmost sorrow," and he became a man of science with the "lively perception" of "the ruination of individual happiness" that his new-found materialism would likely bring about. Romanes attempted to "stifle all belief of the kind which I conceive to be the noblest, and to discipline my intellect with regard to this matter into an attitude of the purest scepticism." In the process, the universe "lost its soul of loveliness."[92] The lament haunted Romanes throughout his short life and, despite his best intentions to stifle an emotional attachment to God and Creation, he failed. He accepted all the principles of a Darwinian creed and adjusted his practices and his morality accordingly, but his inner feelings were always at odds with his actions. Romanes wielded the scalpel and defended the right of other physiologists to do the same, but ultimately he returned to the notion that judgment would come, not from human mastery of material nature but from on High. Huxley had beaten a drum for physiology, as long as he did not have to do it himself. Romanes had been a physiologist in word and deed but, despite his best efforts, never in heart.

5 Sympathy, Liberty, and Compulsion: Vaccination

Sympathy Enforced

As early as 1833, the moral philosopher and prominent Scottish churchman Thomas Chalmers had warned against the state's appropriation of matters of compassion, whether through legislative interference in morality or through the appropriation of the business of charity. "Nothing," he said, "more effectively stifles compassion, or puts it to flight, than to be thus meddled with."[1] The fears represented here remain current in a modified form. In her advocacy for the compassionate state, Martha Nussbaum acknowledges that institutions with the compassion "built in" might be thought of as making individual compassion unnecessary. She counters that a utopia is unlikely, and that compassionate individuals will always be necessary to keep institutions in check.[2] But such views overlook the spiritual content of compassion in historical context—that an institutional compassion was essentially Godless, and that individual compassion was an open relation with a divinely instilled conscience (not put there for the social good, but for the individual's good). Chalmers's fear was that in institutionalizing compassion, a relationship with God, and therefore the basis of morality itself, was being degraded. Such was the theological approach to the limits of government responsibility for moral questions, which survived well into Darwin's later years. It underscored the reaction to a new departure for the state in its assumption of an interventionist public-health role.

Compulsory vaccination in England in the second half of the nineteenth century has received no shortage of historical attention. It is a moment of enormous proportions in the social history of medicine, bringing the institutions of science and medicine into conflict with "ordinary" citizens through the full juridical force of state power. The claims of those at the vanguard of experimental medicine had transformed Jenner's discovery of the protection afforded by cowpox against smallpox in humans from a social good to a social imperative.[3] Despite the absence of a precise understanding of how vaccination worked, statistical evidence was employed to demonstrate its clear effectiveness. To leave a child unvaccinated, given such a weight of evidence, seemed to imply not only neglect of that individual but also a tangible risk that through such unprotected hosts the disease itself would be kept alive.[4]

A series of Vaccination Acts in England established the principle of compulsory medical intervention in the lives of every newborn child. Vaccination against smallpox was made compulsory in Britain from 1853, with the principle of multiple prosecutions for individual parents who refused or neglected to comply from 1867, and with measures effectively to enforce compulsion from 1871. This heralded the beginnings of a systematized national, but organizationally local, health service, the tightening of procedures to do with birth registration, and the closer statistical monitoring of the whole population.[5]

Heavily involved in the institutionalization of both the practice of compulsory vaccination and the government organizations that oversaw it was John Simon, whose dogged axiom of emotionally controlled work for the greater good we have already encountered in relation to the vivisection controversy. Simon does not play a huge part in the analysis in this chapter, but it is important to recognize the continued presence in the early part of these debates of a committed Darwinist who understood the medical need to discard immediately apparent, if not ultimately substantial, ethical objections reached through emotional reactions. The country's first Chief Medical Officer, Simon oversaw the Sanitary Act of 1866 and the Public Health Act of 1875, all the while offering bundles of medical and statistical evidence in support of the efficacy of vaccination, and of the ethical imperative of its compulsory enforcement.[6] He penned in 1857 the *Papers Relating to the History and Practice of Vaccination*, which became the authoritative evidential text in support of the practice, and which was repeatedly utilized in Parliamentary debates on the ethics of compulsion.[7] His outlook on the necessity of enforcing public health for the common good was abundantly clear. His annual report to the Local Government Board in 1874 noted the waste of human life, caused largely by "reckless disseminations of contagion," and blamed local authorities for failing to use "with adequate skill and vigour the resources which are in their option to use against the evil." Simon wanted to light a fire under the authorities to encourage them to use their legal, judicial, and penal powers for the "due protection of human life."[8] Simon counted Darwin among his personal friends, and the connection carried weight in his role as Medical Officer.[9] In his public-health outlook, Simon shared a philosophy with Darwin.

The Vaccination Acts interjected the authority of the state into the private sphere, exposing parents for the first time to the prospect of having no choice with respect to one matter of their offspring's well-being. Judith Rowbotham has pointed out that this was "symptomatic of a new desire to regulate in areas hitherto regarded as secular, and so outside the scope of government moral regulation." With scientific authority emerging as the new (and highly resisted) source of moral authority, resistance "counted as the new heresy."[10] Refusal to submit to the vaccinator's lancet could lead to repeated prosecutions resulting in sizable fines and prison sentences, sometimes with hard labor. In some communities, Leicester being a prime example, resistance to vaccination led to significant social upheaval, with the local M.P. and elected councilors in open defiance of the law, thousands of citizens handed out criminal convictions, vast swathes of the population unvaccinated, and enormous protests in the streets in the name of the liberty of the parent.[11] The resistance was based on a collection of fears mixed with political principles. Some of those fears were based on well-founded concerns about the risks of a nonstandardized and sometimes unhygienic medical procedure; some on rampant paranoia and superstition, the flames of which were fanned by radicals both locally and nationally. The political principles were sometimes well-founded objections to the imposition of state power into the sovereign domain of the home, and of elite abuses in the name of suppressing or "ordering" the working-class population; sometimes they were equally paranoid expressions of suspicion about the vested interests of the institution of medicine and its quacks on the make.[12] The controversy led to a six-year Royal Commission, with thousands of pages of testimony, concerning both the efficacy of vaccination and the ethics of compulsion.[13]

For three decades, vaccination against smallpox remained a prominent ethical debate concerning the state's humanitarian and communitarian role in safeguarding public health. Vaccination was, to many, a callous imposition of state medicine into the private affairs of families, especially after 1871, when parents of newly registered infants were commanded to vaccinate their children within three months of registration or be prosecuted. The decision to vaccinate was not straightforward: vaccination sessions were carried out publicly and administered by Poor Law Guardians, forcing the shame of the workhouse onto those who would otherwise have shunned it, and bringing the degenerate poor literally shoulder to shoulder with those who identified as respectable. Arm-to-arm vaccinations were commonplace, and stories abounded of accidental contamination of syphilis and erysipelas from the unclean poor. Many reported deaths directly or indirectly attributed to vaccination. Moreover, there was not a universal procedure of vaccination, the skills and methods of vaccinators varying greatly throughout the nineteenth century, and with only a hazy understanding of how vaccination worked at all, for how long it lasted, and what its effects on immunity really were. Even where vaccine lymph was not drawn from the vesicle of another arm, but taken directly from the calf, it was nevertheless suspected of impurities and adulteration, leading to disease. These doubts, combined with a righteous indignation at the imposition of state authority against the liberty of the child and of the authority of the parent, led many to dissent. Smallpox vaccination generated fear in urban spaces that rivaled the fear of smallpox itself. But dissention was no easy call either. Having refused to vaccinate a child, a parent (usually the father) appeared in court, and was prosecuted, fined, or imprisoned. Subsequently, a new order to vaccinate was issued, and a further refusal merited a further prosecution, and so on, leading some to destitution through fines and the distrainment of their material goods—the forced entry of bailiffs into private houses and the removal and subsequent public sale of furniture—in lieu of payment.

So much we know. This chapter pursues a little-explored element of the public, scientific, and medical debate by focusing on the specific contribution to it of prominent evolutionary scientists, many of whom we have already met. One might assume, on face value, that such men would have little to no stake in the ethics of compulsory vaccination, but it is my contention that vaccination offers a sturdy test-case of the practical implications of evolutionists' theories of the development of morals. At the heart of the vaccination controversy was the question of whether or not its compulsory infliction was moral.[14] Evolutionary science's development of sophisticated theories of moral development, moral action, and social good thrust the fathers of those theories into the debate. They thought themselves equipped to intervene in this moral quandary with the weight of a fully formed theory to support them. The problem, and the reason this is particularly worthy of note, is that armed with the same intellectual tools, the four principal evolutionists—Charles Darwin, Herbert Spencer, Thomas Henry Huxley, and Alfred Russel Wallace—could scarcely agree. In fact the vaccination scandal would cause irreconcilable divisions. Those divisions ultimately depended on a disparity of interpretations of the efficacy of natural selection among humans and its implications for an understanding of liberty, and, most importantly, on a fundamental disagreement about the practical implications of the evolution of sympathy and moral action.

Because Darwinism had itself raised the specter, among opponents of evolutionary theory, of a degradation of morality in favor of the "survival of the fittest," Darwin's implicit assertion that his evolutionary account could not only explain civilized morality but prescribe its future development is historically important. One discrete lens with which to bring the history of that attempt into focus is the controversy surrounding compulsory vaccination. Darwin, as we shall see, unequivocally supported compulsory vaccination. How can one understand that support, under threat of legal penalty, in the name of sympathy? Or, more broadly, how can one interpret the physical acts of compulsory vaccination—the lancet in the arm, the enforced visit to the workhouse, prosecution in court, a stay in prison, the removal of furniture—as sympathetic? To answer that, one must recall the centrality of sympathy in Darwin's account of the evolution of morality and of civilized society, and thereafter follow what his peers did with it.

Wicked Folly

The Act of 1871 coincided with a smallpox pandemic across Europe and with the publication of Darwin's *Descent of Man*. Clearly, smallpox had been in Darwin's mind as he was finishing the book. Discussing the universal extension of sympathy among "civilised men," working to check the suffering of natural selection, Darwin explicitly gave as one of his examples of this phenomenon the belief "that vaccination has preserved thousands, who from a weak constitution would formerly have succumbed to small-pox."[15] According to John Simon, the "extraordinary storm of smallpox" peaked in London at the beginning of 1871, and "tested to the very utmost the value of the defences which we [. . .] had reared against such attacks." To make matters worse, a Select Committee of the House of Commons was sitting at the beginning of 1871 to consider the efficacy of the Vaccination Acts, at which vaccination's "accusers" had their say. Simon carried the day, establishing the "powerful protective value of vaccination" to his own satisfaction, and reinforcing the "principle of the Act which had made infantine vaccination compulsory."[16] But it did not hurt to have the endorsement of the nation's greatest scientist. For Darwin, the Vaccination Act was an example of highly evolved sympathy, institutionalized in law by eminent men, who by medical science and public influence were putting an end to the suffering that nature would otherwise dispense. Darwin observed how the Santali "savages" or "hill-tribes of India" had become more prolific, increasing "at an extraordinary rate since vaccination has been introduced."[17] It seemed on face value to be a factor in the preservation of the "weak," but perhaps smallpox was no indication of hereditary weakness. It struck at rich and poor alike, and on what basis was the discerning man supposed to decide who to exclude from this privilege of immunity? And should a parent object, on grounds of superstition or stupidity, was it not the duty of the sympathetic community to overrule that objection, for the good of the whole? Darwin was a firm advocate of vaccination, carefully marking the vaccination dates of his own children in the family Bible. It was, he said, socially "beneficent," and he called its opponents "bigots" and supposed that they were "too ignorant to be able to see their own ignorance."[18] According to the recollections of William Darwin, Charles had had "very strong feelings against what he felt to be the almost wicked folly of the anti vaccination agitation."[19] Nevertheless, some still asked about the greater good of society in the long run. Though the legislation enforcing compulsory vaccination may have seemed like an act of highly evolved sympathy—an unavoidable measure of civilized compassion for those most likely to suffer—might it not in the long term be a misdirected and self-defeating piece of class legislation, through which the worst of civilization would be preserved, to the overall detriment of society? While smallpox struck rich and poor alike, it struck the poor in far greater numbers, being spread most quickly in those areas that were most densely populated. Ought not those who would, in the natural run of things, succumb to the disease, be allowed so to succumb?

John Stuart Mill had suggested as much to Herbert Spencer as early as 1861, pointing out that vaccination ensured the survival into adulthood of individuals of a weak disposition, who otherwise would have succumbed to disease in childhood. This explained the "diminished strength of constitution of the middle and higher classes" in comparison to the "strong constitutions of many savage tribes."[20] Spencer, who for his part agreed with Darwin that sympathy lay at the root of the civilized moral sense, was nevertheless opposed in principle to the compulsory enforcement of vaccination (and also had grave doubts as to the medical efficacy of the procedure).[21] Spencer came to see the imposition as certain to prevent the natural evolution of those impulses necessary to self-regard, which was a cardinal marker of his life's work.[22] He was an anti-vaccinationist on the principle of liberty, but his liberty had an evolutionary stamp. Government intervention in public health in general, enforcing medical benefits instead of leaving them to the individual to seek them out, would ultimately adversely affect the population. It denuded individuals of their own survival instincts and preserved the weak. Whereas Darwin opined that civilized man spread his kind through public opinion backed up by the force of law, Spencer pointed out that such laws prevented the development of those higher emotions in the first place.

A man will not be more likely to jump into the river to save a drowning boy because you tell him he must not stand idly by. To abstractly rationalize sympathy—to legislate for it—would likely check its natural development.[23] And just as one ought not legislate for self-sacrifice and risk, one ought not legislate for self-preservation and the avoidance of risk, lest that too check its natural development. As generations passed, the instinct of self-preservation would atrophy, causing an ever-greater reliance on the state to intercede.

The increasing stringency of the Vaccination Acts were, to Spencer, evidence of this trend, and Spencer accused liberals of trampling liberties under foot "in pursuit of what they think popular welfare."[24] He chided the promoters of state-run vaccination with the pregnant threat that "Where once you interfere with the order of Nature there is no knowing where the results will end."[25] Spencer wished to "avoid assisting the incapables and the degraded" in multiplying and resisted what he called the "tyranny of the weak."[26] Whereas for Darwin it was too great a sacrifice of the noble civilized sympathy to overlook the poor health of the unfit, Spencer wanted to let nature take its course. To him, compulsory public-health measures were an example of a misplaced, less-evolved compassion, serving no benefit.

T. H. Huxley disagreed with Spencer about a great many things, and vaccination was not the least of them. The two men were friendly in person, but publically at odds. In 1887 Huxley refused an honorary position in the London Liberty League on the grounds that the same offer had been made to Spencer. "Now you may be sure that I should be glad enough to be associated with you in anything," Huxley told him, "but considering the innumerable battles we have fought over education, vaccination, and so on" the program of any League that would have them both as figure heads "must be so elastic as to verge upon infinite extensibility."[27] Huxley's views can be best understood through the lens of his notion of civilization and sympathy's role in it. Darwin's sketch of the emotional basis of the "primitive bonds of human society," which becomes through evolution "the organized and personified sympathy we call conscience," was relabeled as the "ethical process" by Huxley. But in so formulating it, Huxley departed from Darwin in noting that the Golden Rule can "be obeyed, even partially, only under the protection of a society which repudiates it," pointing out that an absolute observance of the Golden Rule would be "incompatible with the existence of a civil state." After all, Huxley asked, what would happen to a garden "if the gardener treated all the weeds and slugs and birds and trespassers as he would like to be treated, if he were in their place?" In highly civilized societies, according to Huxley, the struggle for the means of existence was at an end, and with it the "ethical process." The appearance in these highly civilized societies of degenerates whose lives were nasty, brutish, and short—the weeds, slugs, birds, and trespassers—was a mere apparition of natural selection. Under the right conditions, which a civilized state had the power and means to institute, there would be no "unfit" among the population. Huxley was able to make an argument for medical policy at the level of the state on the basis of his egalitarianism. For, he said, the "higher the state of civilization, the more completely do the actions of one member of the social body influence all the rest, and the less possible it is for any one man to do a wrong thing without interfering, more or less, with the freedom of all his fellow-citizens." The state's role was therefore significant, for individual liberties could, especially in matters of health, jeopardize the whole. Huxley complained that if "my next-door neighbour [. . .] be allowed to let his children go unvaccinated, he might as well be allowed to leave strychnine lozenges about in the way of mine." It did not mean, as Huxley assumed Spencer to take it to mean, that "the principle which justifies the State in enforcing vaccination or education [. . .] also justifies it in prescribing my religious belief [. . .] or the pattern of my waistcoat." On the contrary, a "properly organised State," the government of which "being nothing but the corporate reason of the community, will soon find out when State interference has been carried far enough." Huxley compared the state to a physician who will not administer drugs if they do more harm than good in the long run, who in "three cases out of four" will do the "wisest thing" and wait, leaving "the case to nature." "But in the fourth case, in which the symptoms are unmistakable, and the cause of the disease distinctly known, prompt remedy saves a life."[28] And so too with the state, which must administer to the patient—society as a whole—for its own good.

The philosophy found important adherents at the vanguard of medical science as well as on the front lines of vaccination enforcement. Joseph Lister, describing the importance of a working relationship between science and medicine before the British Association in 1896, commented favorably on the punitive measures employed to ensure compulsory revaccination in Germany. If the quality of lymph could be guaranteed, Lister saw no "rational basis" to any kind of conscientious objection to the law.[29] Meanwhile, A. F. Vulliamy, clerk to the Ipswich Board of Guardians, summed up the practicalities and the stakes succinctly:

> Surely it is the duty of all, and more especially of those who, as ministers of religion, or as large employers of labour, can influence many minds, to use such power as they possess to educate and impress those who look up to and believe in them, to obtain for themselves and their children this safeguard freely given to all. And if men will remain obstinate in their blindness and prejudice, then

it becomes the duty of those upon whom is cast the enforcement of the law to consider rather the good of the community than the wishes of the individual; and, if need be, to save helpless infants from danger in spite of their parents. If they neglect this duty cast upon them by the legislature; if, for the sake of momentary popularity amongst the ignorant and prejudiced, they refuse to exercise the power given them to confer an inestimable boon upon those who cannot think and act for themselves, and an outbreak of small-pox occurs, as it may any time, the deaths and disfigurement of hundreds, maybe thousands, will lie at their door.[30]

For Huxley, there was no doubt about the efficacy of vaccination in staving off smallpox, and he set about trying to correct common prejudices. Huxley's role in popularizing germ theory led him to extol the virtues of vaccination, from the point of view of a science of contagion, from public platforms. Against anti-vaccinationist claims of the quack science of vaccination, Huxley pointed out to a large audience in Manchester in 1871 that "experiments of the most ingenious kind" had succeeded in separating out the solid, living parts of a disease from its fluid parts, in order to prove that "you may vaccinate a child as much as you like with the fluid parts, but no effect takes place." But "an excessively small portion of the solid particles [. . .] is amply sufficient to give rise to all the phenomena of the cow pock," revealing the "astonishing analogy between the contagion of that healing disease and the contagion of destructive diseases." This science of transmission, which Huxley compared to the action of yeast, would help science alleviate, or wholly prevent, "some of the greatest scourges which afflict the human race."[31]

The engagement with the science of contagion is key, for it meant that vaccination was not an article of faith for Huxley, but an empirically provable phenomenon. There was no need to engage in arguments of comparative statistics (a preoccupation of Wallace, as detailed below) when one could point at a microscope slide. He had taken a similar tack in his presidential speech to the British Association in 1870, but with the important addition of a moral. The Darwinian creed of sympathy was intrinsic to scientific practice, lest "the people perish for lack of knowledge." This was not a matter of choice, but a societal imperative: "the alleviation of the miseries, and the promotion of the welfare, of men must be sought, by those who will not lose their pains, in the diligent, patient, loving study of all the multitudinous aspects of Nature, the results of which constitute exact knowledge, or Science."[32] Anti-vaccinationism may have been in earnest, but Huxley saw it as clearly borne of ignorance.[33] It was not sufficient, therefore, that vaccination be promoted; anti-vaccination had to be forcefully stamped out. To that end, Huxley led a deputation of the Royal College of Physicians, the Royal College of Surgeons, and the Royal Society, to the Local Government Board in 1880 to prevent a change in the law that would have softened the penalties on those parents who refused to vaccinate their children. The proposed Vaccination Amendment Bill was designed to prevent the multiple prosecution of parents for failure to vaccinate, limiting the fine to twenty shillings. The deputation submitted, in particularly Huxleyan terms, that such an Act would be "at once a premium held out to ignorant people to evade the law by the payment of a small penalty, and that small-pox might easily be thereby spread amongst the people, and a visitation similar to that of 20 years ago might ensue, which destroyed the lives of thousands of people." Urging the "efficacy of vaccination as a protection to the people," the deputation predicted the "fatal consequences" that would follow the passing of the measure.[34]

Huxley's own deputation had been preceded by one from the British Medical Association, which had included the noted Darwinist Sir John Lubbock, who observed the "vast preponderance of opinion" in favor of vaccination.[35] That deputation also rehearsed Darwin's moral creed in an exemplary manner. Ernest Hart, one of the foremost medical journalists of the age, as editor of the *British Medical Journal*, put the case thus:

> [I]f the health of the population required that any members of it, and especially the infant members of it, should not be left to become sources of danger to the whole nation, they [the medical profession] thought that political expediency required that they, their guardians, should be prevented from allowing them to become sources of danger. If a man could have small-pox all to himself, he might fairly claim to be allowed that privilege, and the State need not interfere; but he could not do so, unless he consented to quite exceptional quarantine arrangements for himself and family, which, in our complicated society, were practically impossible.[36]

The somewhat extraordinary assertions that the medical profession policed the guardians of the infant population, that the institution of medicine had something to say about political expediency, and that the prevention of disease qualified doctors to define the limits of individual liberty, were all in accord with Darwin's evolutionary account that behooved medical science's far-seeing sympathy to act for the good of civilization. Those who believed otherwise, at least as far as Huxley was concerned, were no more credible than the advocates of spiritualism.[37] Alfred Russel Wallace, the codiscoverer of the process of evolution by natural selection, prominent apologist for spiritualism, and anti-vaccinationist *par excellence,* doubtless took note.

Wallace Stakes His Reputation

By the time the vaccination controversy peaked, Alfred Russel Wallace's ethics and evolutionary reasoning were far from Darwin's, and Wallace's own position within the scientific community was mocked by its leading lights, even if his popularity did not wane.[38] Wallace argued for the exceptionalism of the human mind and advocated the exploration of spiritualism as a science. To complement this, he took up a radical political position over land reform, and argued passionately against vaccination.[39]

Wallace's anti-vaccinationism can be viewed as a composite of the evolutionary ethics of Huxley and the politics of Spencer, with his own statistical evidence concerning the medical efficacy of vaccination superadded. Wallace was clear that natural selection had ceased to act among civilized races because of humanity's "superior sympathetic and moral feelings," which had fitted it "for the social state." Under these conditions, he said, man

> ceases to plunder the weak and helpless of his tribe; he shares the game which he has caught with less active or less fortunate hunters, or exchanges it for weapons which even the weak or the deformed can fashion; he saves the sick and wounded from death; and thus the power which leads to the rigid destruction of all animals who cannot in every respect help themselves, is prevented from acting on him.[40]

He, like Huxley, exempted civilized man from the vagaries of natural selection. This being the case, a man could, or possibly should, go out of his way to save those who seemed weak, for weakness was not, among the civilized, that potent marker of annihilation as it was in Nature. Wallace's argument about vaccination was that it did not alleviate suffering, either in terms of medical efficacy or in terms of distribution. In fact, it made suffering worse. It was therefore was a corruption of civilization.

Turning the tables on Huxley, he accused legislators of ignorance, in thrall to the medical establishment, who, in turn, were interested primarily in lining their own pockets. As far as Wallace was concerned, "lymph"—the matter taken from a cowpox pustule and injected into children—was a misnomer. It was actually "diseased and disease-bearing matter, and should be called pus, and its willful insertion into the skin of any human being should be called blood-poisoning, and denounced as a crime of the first magnitude." Only through such a denunciation could it be brought home to "the average legislator" that he was responsible for "the terrible consequences of his ignorance and his submission to a prejudiced and interested profession."[41]

Wallace was convinced that the portion of the population who remained unvaccinated were largely the "very lowest class, including the tramps and the criminal classes, and also children who were too delicate to be vaccinated, and children who got small-pox before they were vaccinated." In short, the poor, including what he called the "very dregs of the population," were much less vaccinated than the rich.[42] Poverty, ill-health, criminality, and weakness *per se* were more likely, Wallace thought (despite evidence to the contrary), to make a person susceptible to smallpox when compared with the rich and the strong. Vaccination statistics were therefore useless, since the fact of being vaccinated was not substantial proof that this caused immunity, compared with the fact of being rich. Indeed, so far as Wallace was concerned, the statistics proved that the struggle for existence was being artificially preserved by a policy of class interest that falsely flew under the banner of public health. Epidemics themselves caused immunity afterward "by simply killing off the susceptible people."[43] The "community is divided between those susceptible of small-pox and those not susceptible," Wallace said. "[I]n a bad epidemic almost all the susceptible have it and get over it, or have it and die."[44] Vaccination, therefore, was nothing more than "a medical dogma [. . .] enforced by penal law," a "cruel and tyrannical" intervention with "no beneficial effect whatever," which actually stood in the way of the kind of sympathy that was supposed to mark a society in a state of civilization.[45]

Vaccinators were thus social exploiters, enriching themselves at the expense of the poor. Wallace's anti-vaccinationism was clearly shot through with his socialism. His belief in the power of evolved sympathy among fellow men was not to be extended to a blind faith in the peddlers and institutions of social injustice from on high. Insofar as Darwin had refined Adam Smith in the mode of *The Theory of Moral Sentiments,* Wallace now channeled Smith in the mode of *The Wealth of Nations*. To expect "exploiters" to be converted into altruists was "a fatuous programme—a maniac's dream." Only when the "exploited" acquired "enlightened self-interest" would social injustice be cured. Wallace hoped for the exploited to be "equally [. . .] consistently, persistently egoistic" to counter the egoism of the exploiters. Egoism would spell "justice and freedom, as surely as altruism spells charity and slavery":

> This I take to mean that we must be guided by egoism—enlightened self-interest—in all relations between Socialists and non-Socialists, while *not* implying that we are to be egoistic in our individual relations with our fellow-men; that we are fools if we choose landlords, capitalists, soldiers, or lawyers to represent and govern us; and that the kind-hearted and altruistic workers and dispensers of charity in the slums, though individually to be admired

and loved, are as a class to be condemned, because they have no object but to palliate symptoms instead of removing causes.[46]

In his evidence before the Royal Commission on Vaccination, Wallace was explicit that compulsory vaccination was part of this scheme of murderous exploitation.[47] He was even more candid in his communications to the National Anti-Vaccination League, proposing that "Our legislators should be told at every possible opportunity that, by permitting the Vaccination laws to continue on the statute books, they are responsible for the deterioration of the race, for untold agony, physical and mental, and for countless legalised though officially-concealed murders."[48] Given what he saw as the pure politics of vaccination policy, Wallace felt compelled to intervene because, he said, "Liberty is in my mind a far greater and more important thing than science."[49] Communal sympathy may have been the basis for the rise of civilization, but the maintenance of artificial hierarchies threatened social welfare and preserved a sham of natural selection that could only be demolished by the asserted self-interest of the working man.

The only problem, which caused Wallace some difficulty before the Royal Commission, was that the "control" town of Leicester, whose population refused *en masse* to be vaccinated, relied for its protection on an even greater imposition of external authority. In the event of a case of smallpox, not only the victim, but also all the family members and contacts of the victim, were compelled to isolation beyond the confines of the city. Even more uncomfortably for Wallace, and to the great delight of the medical community, this relied for its effectiveness on all the medical staff in the isolation hospitals being revaccinated. Leicester was an unvaccinated island bound by a vaccinated cordon.[50]

Wallace knew that his stance on vaccination was alienating him from the world of science, but he stuck to it because he thought it "about the most demonstrative bit of work I have done."[51] He also saw that his eminence in the public world at large might bring some influence to bear, particularly with men who shared some of his other interests. Frederic William Henry Myers, for example, was an exponent of psychical research into the paranormal, but he also had a role as H.M. Inspector of Schools. He was strongly pro-vaccination but told Wallace that "if I am converted it will be wholly *your* doing," adding "there is *no one* by whom I would more willingly be converted than yourself."[52] Wallace sent his 1898 pamphlet, *Vaccination A Delusion, Its Enforcement A Crime*, to friends, colleagues, acquaintances, and fellow agitators in the hope of convincing them, statistically, that vaccination had "been the cause of so much disease, so many deaths, such a vast amount of utterly needless and altogether underserved suffering, that it will be classed by the coming generation among the greatest errors of an ignorant and prejudiced age, and its penal enforcement the foulest blot on the generally beneficent course of legislation during our century."[53] The sympathetic course, in evolutionary terms, was to smash the corrupt institution of professional medicine, which Wallace saw, as Huxley had prophesied, as a "black art."[54]

Of course, Wallace's broadsides were an abject failure, his converts few, and the proponents of professional medicine were unsparing in their criticism. *The Lancet* gave the pamphlet short shrift, referring back to Wallace's reluctance before the Royal Commission to entertain other statistical sources that went against the tenor of his own. It is perhaps a fitting irony, given that evolutionary scientists' engagement in this topic centered around the meanings and applications of liberty and sympathy, that *The Lancet* pointed to the fact that such statistical selectivity "could hardly be expected to meet with much sympathy," lamenting that his performance before the commission, and his corresponding pamphlet, were not worthy "of his great reputation."[55]

Conclusion

The four names most prominently associated with the theory of evolution in the second half of the nineteenth century all shared a conviction about the rule of natural law. They also shared a conviction about the evolution of sympathy and its central role in the development of moral societies, that is, civilization. Darwin's theory of natural selection was at the center of all four men's understanding of how natural law worked. The men diverged, however, when it came to their respective understandings of the functioning of natural law in civilized societies. Both Darwin and Spencer were convinced of its continued importance. Darwin saw how the "domestication" of society could alter the working process of selection, and Spencer lamented any such artificial means of preserving those who otherwise would be found to be "unfit." Huxley and Wallace thought that civilization brought natural processes of selection to an end. This divergence had profound implications for what constituted sympathy and, accordingly, moral action in civilized societies. It was also significant for their respective understandings of the role of the state in guaranteeing or limiting liberty, especially in the name of public health. Where Darwin and Spencer agreed about natural selection, they clearly departed from each other on what to do with the "weak"; and where Huxley and Wallace agreed about the cessation of natural selection, Wallace's political and statistical bents led him to break sharply from Huxley about the question of liberty and the state.

Their varied interpretations led them to distinct political outlooks regarding the ethics of state compulsion of a public-health measure. The morality of the action of administering cowpox lymph to a child under the threat of penal action against his or her parents was judged according to the evolutionary role of sympathy within civilized society. As these men brought their own highly controversial moral compasses to bear upon a scientific scandal that seemed at once to usher in the principle of social medicine and to demolish the principle of individual liberty, they helped fan the flames of outrage in the medical establishment and of fear in the population at large. Though it would be difficult to substantiate any claim that any of these men had an instrumental effect on how the vaccination policy was pursued, enforced, or ultimately relaxed, it is surely of great moment that such giants of science lent their not inconsiderable force to both sides of the public debate, helping to drive opinion about a state policy of medical intervention in the name of an evolutionary account of morality.

The questions of selection, domestication, and racial fitness were, in and of themselves, drivers of social policy, social investigation, and public opinion. Where evolutionary morality had been brought to bear upon the vaccine debate that had been running since the very beginning of the nineteenth century, it was also at the heart of novel social and political movements spearheaded by more faithful followers of Darwin. Where public-health policy pondered what to do with those who were already alive, the new science of eugenics concerned itself with the politics of breeding. At the heart of this new science lay Darwinian sympathy, practically repurposed for social, scientific, and ultimately moral ends.

6 Sympathetic Selection: Eugenics

When Francis Galton read the masterwork of his cousin, Charles Darwin, in 1859, it altered the course of his own life and work.[1] *Origin of Species* was scrupulous in its explicit avoidance of developing the theory of natural selection to humans, saving that hammer blow for the 1871 sequel, *Descent of Man*. Regardless of the omission, the pieces were easy enough to put together. It was Darwin's major essay on the domesticated breeding of fancy pigeons that most struck Galton. Darwin had told the story of natural selection by analogy, pointing out how easily man could contrive this process of selection under controlled conditions, breeding for those qualities he desired and eliminating those he did not. Darwin sought principally to describe these things. His analogy of nature was only that. Galton saw the possibilities of extending the principle of controlled breeding to humans. If natural selection had brought about civilization, the scientific insight that had rendered this manifest also held the key to bringing an end to the role of nature in selection. Galton would dedicate the rest of his life to the promotion of trumping nature at its own work, finding a science of selecting for the fittest and eliminating the weakest among humans.

I am going to put the early theoretical work of Galton and his followers, notably Karl Pearson, on eugenics into the context of their blueprint of the evolution of civilization, their personal relationships, and their ideas about the moral character of the scientist. Central to both their theoretical approach to human breeding and to their construction of the scientific self was a notion of sympathy that could both rationalize the elimination of the "less fit" and qualify "scientific men" to lead society in the project of eugenic betterment. This unfamiliar sympathy marked out a way of ordering and regulating society, and prescribed the civilized individual, in whom the intellectual and the emotional would be conjoined.

I do not intend to offer a general reappraisal of the history of eugenics *per se*. That historiography is already large and still growing.[2] Moreover eugenic ideas were, as Marius Turda has put it, "a polysemic system of thought."[3] No one characterization can sum up the movement or its meanings, and this chapter should not be read as a reductionist attempt to reappraise the historiography. Rather, it is an attempt to introduce the history of emotions into a field of study where it has so far been largely absent.[4] It argues in particular that the redefinition of sympathy in Darwinian concerns was an essential precondition for Galton's particular brand of eugenics. Furthermore, Galton's appropriation of this new sympathy, turning it to his own ends, was fundamental to his interpretation of what was wrong with civilized society, and how to fix it. The history-of-emotions perspective, which encapsulates Galton's and Pearson's own understanding of the possibility for and advantages of emotional change, introduces a new and important causal factor into at least this strand of the history of eugenics. It is therefore important, first of all, to put the work of Galton and his disciples into the context of Darwin's skepticism about the extension of the idea of controlled selection to humans. That skepticism was not borne of any doubt that a science of breeding could be extended to humans. Whatever his reticence to discuss selection among humans in *Origin of Species*, his later work (which had in turn been influenced by Galton in the intervening years) went into great detail on the subject.

Darwin's Paradox

The argument of this book has been building to this point. We have already seen, especially in the arguments among evolutionary scientists over vaccination, that there was significant disagreement about the moral imperatives arising from Darwin's theories. What to do for the best, for the overall good of society, depended on whether or not one subscribed to Darwin's view only up to the point at which he forecast the universal extension of sympathy to all nations, races, and creatures. What follows that forecast, that ever-progressive cycle in *Descent of Man*, is a sudden and paradoxical degeneration.

Civilization contained the seeds of its own demise. Darwin was perfectly aware of the problem. "With savages," he said, "the weak in body or mind are soon eliminated; and those who survive commonly exhibit a vigorous state of health. We civilised men, on the other hand, do our utmost to check the process of elimination; we build asylums for the imbecile, the maimed, and the sick; we institute poor laws; and our medical men exert their utmost skill to save the life of every one to the last moment." Darwin ascribed the impulse of the civilized to "give to the helpless" to an "incidental result of the instinct of sympathy, which was originally acquired as part of the social instincts." And what is more, this impulse could not be switched off without the deterioration of what Darwin called "the noblest part of our nature." For if, Darwin said, "we were intentionally to neglect the weak and helpless, it could only be for a contingent benefit, with an overwhelming present evil." Darwin foresaw the inevitable result, that "the weak members of civilised societies propagate their kind." It caused him to lament that "excepting in the case of man himself, hardly anyone is so ignorant as to allow the worst animals to breed." For Darwin, at least, there was no choice but to resign himself to preserving the weak and to hoping they would not procreate.[5]

But where Darwin clung to the nobility of humanity in preserving the weak, his account of the evolution of sympathy and of civilized morality sounded a different note to others.[6] Herbert Spencer, for example, was the first to establish a kind of collision course between medical science and public health policy and the evolutionary progress of the race. He did not formulate a eugenic strategy, but he helped prepare the way for it. As has already been noted, Spencer was actively opposed to the "tyranny of the weak" that would arise from any form of assistance given to degenerates to swell their own numbers.[7] If natural selection was to be the defining rule for civilization, then it should be left to its course. To Spencer, compulsory public-health measures were an example of a misplaced, less evolved sympathy, serving no benefit. He cautioned against charity out of place. For, he said, "If sympathy prompts an equal attention to the improvident as to the provident, the sentiment of justice puts a veto." If, he went on, "as much sacrifice is made for the sick good-for-nothing as is made for the sick good-for-something, there is abolished one of those distinctions between the results of good and bad conduct which all should strive to maintain."[8] To preserve the weak, for Spencer, was to luxuriate in pity, an emotion he described as being excited always by "relative weakness or helplessness." Simply put, "pity implies [. . .] the representation of pain, sensational or emotional, experienced by another; and its function as so constituted, appears to be merely that of preventing the infliction of pain, or prompting efforts to assuage pain when it has been inflicted." Spencer drew attention to a "certain phase of pity" in which "the pain has a pleasurable accompaniment; and the pleasurable pain, or painful pleasure, continues even where nothing is done, or can be done, towards mitigating the suffering," or even when there is no actual suffering at all. Linking this tendency to the "parental instinct," which in Spencer tends to indicate the "maternal instinct," he asked what was the "common trait of the objects which excite" the feeling. It was a feeling that arose "in presence of something feeble and dependent to be taken care of," and could be seen in "the little girl with her doll, in the lady with her lap-dog, in the cat that has adopted a puppy, and in the hen that is anxious about the ducklings she has hatched." Naturally, this extended to "weakly creatures in general, and creatures that have been made weakly by accident, disease, or by ill-treatment."[9] This feeling, a tender sympathy, was a self-serving pleasure, compassion *de haut en bas,* that did not serve any far-reaching good.[10] It accounted for what Gertrude Himmelfarb has called "the corrupt version of the gift as practiced by a lady bountiful."[11] Spencer called this "ego-altruism."[12] Common compassion was selfish.

New knowledge of the natural causes of the moral sentiments would bring this to an end, and "call in question the authority of those ego-altruistic sentiments which once ruled unchallenged." The moral sentiments, once fully evolved, were to "prompt resistance to laws that do not fulfil the conception of justice, [and] encourage men to brave the frowns of their fellows by pursuing a course at variance with customs that are perceived to be socially injurious."[13] With such a vision, the preservation of the weak seemed to Spencer more clearly to be an immoral act because it adversely affected the welfare of the whole society. To be genuinely sympathetic—that is, to have fellow-feeling with other men—was to act in such a way as to reduce the aggregate of suffering, and not enhance it. To eliminate what Spencer could readily identify as the "weak" in society—although Spencer was not sure how to do it—seemed like the *moral* thing to do. The process would have to be begun by those men who deemed themselves more "fully evolved" than the mass.

Galton's Platonic Creed

Francis Galton thought he was such a man.[14] The interests of Francis Galton lay chiefly in the question of heredity, and Galton coined the word "eugenics" to emphasize, literally, the breeding of goodness in human stock. That key term, "eugenics," was in use from 1883. But Galton had been working hard on the principle for two decades by that point. In his first major work, which attempted to isolate why genius tended to run in families (his own genius and that of the Darwins featured heavily), Galton pronounced his opposition to the tendency among civilized peoples increasingly to delay marriage. He protested against the "abler races being encouraged to withdraw in this way from the struggle for existence." From the first, Galton was aware that his ideas might be unpalatable to those whose charitable instincts led them to sustain the "weak" around them. "It may seem monstrous that the weak should be crowded out by the strong," he said, "but it is still more monstrous that the races best fitted to play their part on the stage of life, should be crowded out by the incompetent, the ailing, and the desponding." It was clear even at this stage of his thinking that Galton was thinking in terms of domestication. Just as horticulturists selected for the best seeds in plants, so "viriculturists" would select for the best in humans. "The population of the earth shall be kept as strictly within the bounds of number and suitability of race, as the sheep on a well-ordered moor or the plants in an orchard-house." The science that would make this possible was, in 1869, lacking. But knowledge of its possibility behooved people in the meantime to exercise moral caution, avoiding the "mistaken instinct of giving support to the weak," that negative consequence of the work of sympathy, or common compassion, and instead encouraging the "multiplication of the races best fitted to invent and conform to a high and generous civilization."[15] Following Darwin's idea about the expansion of sympathy to include all the members of society, Galton imagined a civilization in which "personal property is not the foundation," where "republican and co-operative" citizens would act on a "vivid desire" for the "good of the community." He thought evolution tended in that direction anyway, but began to think of it as the "religious duty" of "intelligent men" to "advance in the direction whither Nature is determined they shall go."[16]

Against those who might have objected that society depended on the weak to preserve and cultivate the "pitying and self-denying virtues," Galton was unapologetic. There was no threat of callousness in the breeding of a fit race. It did not seem reasonable to Galton to "preserve sickly breeds for the sole purpose of tending them, as the breed of foxes is preserved solely for sport and its attendant advantages." However fit and healthy the population, Galton was realistic enough to admit that "misery" would never "cease from the land," thereby providing the "compassionate" with "objects for their compassion." Looking around him in the 1870s, Galton only saw endless such objects, which "overstocked and overburdened" the land with "the listless and incapable."[17] Imagining himself in the pulpit of the future, Galton saw the necessity of forcing a change in the emotional disposition of society:

> It is no absurdity to expect, that it may hereafter be preached, that while helpfulness to the weak, and sympathy with the suffering, is the natural form of outpouring of a merciful and kindly heart, yet that the highest action of all is to provide a vigorous, national life, and that one practical and effective way in which individuals of feeble constitutions can show mercy to their kind is by celibacy, lest they should bring beings into existence whose race is predoomed to destruction by the laws of nature.[18]

A refinement of compassion was in order.

As his eugenic ideas matured, Galton became more emphatic that eugenic ideas ought to be "introduced into the national conscience, like a new religion," constructing a new morality that each individual would take on as a patriotic "duty."[19] The prevailing religious morality had fostered exactly the opposite of what a civilized society needed in order to progress. The Church had "brutalized human nature by her system of celibacy applied to the gentle" and "demoralised it by her system of persecution of the intelligent, the sincere, and the free." Galton's blood boiled when he thought of the "blind folly" of the Church in causing "the foremost nations of struggling humanity to be the heirs of such hateful ancestry," which had bred instincts antagonistic to "the essential requirements of a steadily advancing civilization." The internalized sense of immorality and sin had to be overhauled if the pathetic existence of "barren religious sentimentalism and gross materialistic habitudes" were to be overcome.[20] To achieve this required a change in "mental attitude," in which work for the common good became conscious. This was the "religious significance of the doctrine of evolution," which imposed a "moral duty." Who better to lead such a change than the people who thought they knew best how to bring it about? Galton hoped that "a sort of scientific priesthood" could be established throughout Britain, "whose high duties would have reference to the health and well-being of the nation in its broadest sense, and whose emoluments and social position would be made commensurate with the importance and variety of their functions."[21]

According to Karl Pearson, Galton's disciple, Galton was "the first to grasp that if evolution be the true doctrine of the development of living forms, then it is desirable for rational man to take stock of his varieties, mental and physical, to measure their evolutionary value, and to throw himself into sympathy with the changes Nature foreshadows for his kind."[22] This would be brought about through "a new religion, a religion which should not depend on revelation." In Pearson's words, "Man was to study the purpose of the universe in its past evolution, and by working to the same end, he was to make its progress less slow and less painful in the future."[23]

How was this to work? Galton understood Darwinian sympathy to work largely unconsciously: "Life in general," he said, "may be looked upon as a republic where the individuals are for the most part unconscious that while they are working for themselves they are also working for the public good."[24] Such had been the natural evolution of society. Civilized social instincts—human sympathy—had seen the progression of the race as a kind of commonwealth. But left to nature, this evolutionary process was slow and painful. It had been achieved with "great waste of opportunity and life." Measured by his own standards of "intelligence and mercy," evolution had been allowed to run with unnecessary pain and without ruth. Man, "this new animal," was equipped to intervene with his "power, intelligence, and kindly feeling" to reduce the pain but increase the rate of evolution. Not only could this be done, but Galton thought it *should* be done, as man's "religious duty."[25]

Even as early in Galton's eugenic thought as 1883, one can see the beginnings of what I would call the defamiliarization or disruption of the prevailing understanding and practice of sympathy. For it is in the name of the reduction of pain and suffering, and in the name of the welfare of one's fellow men, that Galton pushed for a eugenic morality. Whatever changes to the race might be brought about by the intervention of human contrivance, the intelligence of man would make them run smoothly, and "his kindly sympathy will urge him to effect them mercifully."[26] Eugenics was to be guided by sympathy construed as sympathy for the whole of society. This was building to a mathematical or statistical point. Eugenic sympathy could not function as "mere emotion."

The successful rearing of a master race required detached social engineering. Suffering and sympathy were to become variables in an equation. Without measuring, there could be no true application of sympathy. Here was to be an *evolution* of the emotions, functioning on a higher plane, subject to the full check of highly developed intellect, and with the benefit of society as a whole always consciously in mind. Galton gradually developed his understanding of how human emotions not only evolved, but also how they could be influenced to alter internalized conceptions of moral imperatives. At his most lyrical, Galton talked of the stage upon which "human action" takes place as "a superstructure into which human emotion enters," and on which humans are guided by emotional assurances (as opposed to scientific certainties or statistical probabilities). This "contented attitude of mind" was "largely dependent on custom, prejudice, or other unreasonable influences." We might simply call it culture. Galton saw clearly that it was a cultural war, or a war against culture that he would have to fight, and that it was culture he would have to use to make eugenics a successful creed. He was, like Marx, waiting for the right moment: "When the desired fullness of information shall have been acquired then, and not till then, will be the fit moment to proclaim 'Jehad' or Holy War against customs and prejudices that impair the physical and moral qualities of our race."[27]

Significantly, Galton appropriated the idea of human nobility, that concept which for Darwin stood for the promotion of public health and the preservation of the weak, and put it to eugenic ends. For while Galton's eugenics, in his own words, forbade "all forms of sentimental charity that are harmful to the race," it nevertheless brought "the tie of kinship into prominence" and strongly encouraged "love and interest in family and race." In brief, Galton asserted, "eugenics is a virile creed, full of hopefulness, and appealing to many of the noblest feelings of our nature."[28] Indeed, the original formulation of the term "eugenics" was, by Galton's definition, taken to mean "good in stock, hereditarily endowed with noble qualities."[29] If there was a charitable instinct in society, it was "not unlimited," and spending it on the weak only hindered progress. Charity, contrary to Christian practice and received notions of compassion, was to be "distributed as to favour the best-adapted races." It was not necessary, under such conditions, to pursue the "repression of the rest" because "it would ensue indirectly as a matter of course" if the weak were denied the means of sympathy that they did not merit.[30] He later encouraged "charitably disposed persons" to leave "substantial sums of money to the furtherance of Eugenic Study and practice," instead of wasting it on the alleviation of suffering.[31] Galton did precisely this, bequeathing thousands of pounds for the establishment of the first Eugenics Laboratory, which would be headed up by Pearson.

Galton did occasionally offer practical guidance on this matter of course. Though "valuable citizens" would be encouraged and helped to nurture families and provided with a "larger measure" of "sanitation, of food," and so on, the "sick, the feeble, or the unfortunate" were not to be neglected. But while Galton proposed to do all he could for their "comfort and happiness," the price would be their "isolation, or some other less drastic yet adequate measure" that would effectively stop "the production of families of children likely to include degenerates."[32] Meanwhile, Galton proposed to build up, through research, publication, and influence, "a sentiment of caste among those who are naturally gifted." This "caste" would be bound together "by a variety of material and social interests" and would be taught "faith in their future." The "sentiment of caste" would ensure that they intermarried among themselves.[33] They would easily identify each other by their ability to produce a "eugenic certificate," given out only after the strictest tests of physical, mental, and ancestral health.[34]

The advocate of a positive constructions of eugenics, that is, a system that would encourage the breeding of the most fit, rather than punitively restricting the unfit from breeding, echoed Galton's position. Caleb Saleeby, who was instrumental in founding the populist Eugenics Education Society, and who had a number of public disagreements with the expressly "scientific" Eugenics Laboratory of Karl Pearson, was convinced that sympathy held the key to successful breeding. What he called the "tender emotion" was the thing that "made and makes everything that is good in the individual, and in human society." It was the "basis of all morality," and the reason to "hope for the future of the race." Because "One man, even one woman, is much more sympathetic than another," it was essential to discover a way to measure and locate those potential parents who had the largest measure. Since "hardness and tenderness" ran in families, according to Saleeby's observations, parenthood should become the "most self-conscious thing in life, so that there should be children born to those who love children, and only to those who love children."[35] Cultivating that love, in an era where belief in the inheritance of acquired characteristics persisted, formed part of a general concern with educating the emotions, characters, and morals of the young, that they might break the degenerative cycle that vicious or inebriated parents might have begun.[36] At least Saleeby was optimistic that the cycle *could* be broken, in contradistinction to Galton and especially Pearson.

Galton's plans found their full elaboration in a novel, *The Eugenic College of Kantsaywhere*, that he ultimately aborted after it was rejected for publication. Embarrassed by its literary reception, Galton ditched the project and the manuscript was largely burned. Thinking that certain of the pages might be of value to Galton's protégé, Karl Pearson, the family saved what they thought was relevant and passed it on. Pearson eventually published the fragments of *Kantsaywhere* in his memoir of Galton. We would probably read it as a fascistic nightmare of forced labor camps, forced emigration, and the seclusion, or enforced celibacy, of the mentally and physically ill. Galton envisaged it as an ideal polity, his Platonic republic, designed to cement his "influence on generations to come."[37] It depicts a society by nature xenophobic until immigrants can prove their eugenic worth. In this fantasy, the "fit" population subscribe to an ancestor cult or spirit world, which defines their religious rules of social sin and social good, and which supplies the grounds for who may breed with whom. The grand lie of a eugenic religion is central to social cohesion and stability. The importance of a science of breeding could only be inculcated in the masses through alienation: to breed well was to be a religious duty owed to the state. The novel fragments are extraordinary. They make for turgid literature, but, as Pearson understood, their value was in clarifying the thoughts "of all of us regarding what a society organised eugenically should strive to achieve."[38] At around the same time as his novel was being rejected for publication, Galton asserted that eugenic scientists, those who fed the new creed to the people, were to "justly claim to be true philanthropists." Their enthusiasm was to be "far nobler and more patriotic" than that of the so-called philanthropists of the day. In his memoirs he wrote of the perfect balance of concern with society as a whole and affections for individuals. "Charity refers to the individual; Statesmanship to the nation; Eugenics cares for both." Its precise aim was to train the qualities with which humans were "gifted," namely "pity and other kindly feelings" and the "power of preventing many kinds of suffering." These capacities were to be honed so that natural selection could be replaced "by other processes that are more merciful and not less effective."[39] The eugenic religion, organized around positive incentives for the best to breed and negative incentives (at best) for the worst to abstain, was, according to this definition, a true expression of love for mankind.[40]

Karl Pearson: The First Apostle

The formulation of this new emotional and moral culture was refined and vigorously promoted by Karl Pearson, who had followed Galton from the very beginning of his eugenic ideas. Pearson was chiefly responsible for trying to make a statistical and biometric science out of Darwinian sympathy. He brought together a reverent worship of Darwin, adulation for Galton, a very considerable skill in statistics, a tub-thumping socialism, and a patriotic and imperialistic fervor, to come up with a blueprint for a racially configured socialist nationalism, as a kind of religion. At the heart of all this was an evolutionary understanding of sympathy and a model of what had to happen to sympathy as it had been known in order for the British race to remain pre-eminent in the global struggle of races.[41] Pearson took the view that he was a personal representative of the possibilities of a higher plane of emotional experience, filtered through scientific practice and the cultivation of the intellect.[42] As such, Pearson condemned "the old method of approaching social problems," which was to allow "our sympathetic instincts full play." No longer should "popular feeling" be "stirred by piteous descriptions of wrong and suffering," he said. For this had the effect of preserving and propagating the unfit, against which Pearson demanded an "elaborate study before legislative remedies are sought for social ills." It would take, he said, "scientific knowledge to control our blind social instincts."[43] Sympathy, Pearson warned, "is there ready to run riot in a thousand ways, which sober reflection may not show to be for the ultimate advantage of the herd," and he cautioned that "no judgement will lead to lasting social gain which is reached by appeal to the emotions." The instincts that humans shared with animals had gone astray, leading otherwise right-thinking people to give to beggars or to prop up the blind, deaf, and dumb. This, in Pearson's view, was not conducive to the "total enjoyment of the race." What we needed instead, he said, was "*academic* judgement."[44] This was not a call to annihilate sympathy, but rather to place it on a footing of calculation. Eugenics entailed, he said, "not the repression of the human emotions, but the examination of life from a different [. . .] aspect." With an acute Darwinistic reduction, Pearson declared that the "welfare of humanity" depended on "the destruction of the less fit." Such destruction was a "chief cause of the mental and physical growth of mankind in the past," and, according to Pearson, was the "source of all that we value in the intellect and physique of the highest type of mankind to-day."[45] Knowledge of the ravages of nature's past had armed the scientists of the immediate future with weapons of knowledge that would refine the process of human improvement.

Pearson, faithful Darwinist that he was, acknowledged that sympathy was the result of the struggle "of race against race, and of man against man."[46] After all, he said, "A community not knit together by strong social instincts, by sympathy between man and man, and class and class, cannot face the external contest, the competition with other nations, by peace or by war, for the raw material of production and for its food supply." There was, he said, a "mournful side" of this "struggle of tribe with tribe, and nation with nation," but he thought it "idle to condemn it," for he could "see as a result of it the gradual progress of mankind to higher intellectual and physical efficiency." Pearson exhorted his readers to recognize that the chief gains of this struggle were "civilization and social sympathy."[47] In other words, eugenic calculation would not only subsume sympathy under mathematical practice for the social engineer, but also would effectively enhance the quality of sympathy on a social level, for everyone who survived the social engineer's contrivances. "[W]ider human sympathies, intenser social instincts, keener pity, and clearer principles of conduct:" all these, he said, depended on the "the wreck of nations." Pearson could see "traces [. . .] everywhere [. . .] of the hecatombs of inferior races, and of victims who found not the narrow way to greater perfections. Yet these dead peoples are," he continued "in very truth, the stepping-stones on which mankind has arisen to the higher intellectual and deeper emotional life of to-day."[48] It was only through the "principle of the survival of the fittest, describing [. . .] the continual struggle of individuals, of societies of civilization and barbarism" that science, with Pearson its spokesman, could account for, in his own words, the "origin of those purely human faculties of healthy activity, of sympathy, of love, and of social action which men value as their chief heritage."[49] Put simply, racial warfare resulted in the human capacity for purer love and deeper, more effective sympathy. Making sure that the nation optimized its own racial specimens in order successfully to prosecute these wars became a moral imperative. The mathematician therefore naturally became the moral leader of society.

For the academic judgment of the statistician to work, existing institutionalized structures of sympathy had to be switched off. As Pearson put it, "social sympathy and State aid must not be carried so far within the community that the intellectually and physically weaker stocks multiply at the same rate as the better stocks."[50] Racial integrity within a nation was essential if that nation was to meet the challenge of other nations, and that entailed a properly controlled social instinct, an intellectually and academically directed sympathy. To that end, Pearson despised the charitable organizations he saw around him, and condemned medical science and public health initiatives for leading to the "survival of the unfit."[51] As Huxley had predicted, this eugenicist saw the progress of medicine as a kind of "black art" that would not be fit for purpose until a new class of "medical mathematicians" replete with eugenic morals could be employed to interpret proposed legislation in the name of public health.[52] Again, he did not demand the eradication of sympathy, because it was "absolutely needful for race survival," but he did demand that "all sympathy and charity shall be organized and guided into paths where they will promote racial efficiency, and not lead us straight towards national shipwreck."[53] He even lauded what he called "Our highly developed human sympathy," that led to the indiscriminate relief of "pain and suffering," but only when a new social contract could ensure that the alleviated unfit were barred from breeding.[54] For if humanity was to rise further it would depend, as it had always depended in Pearson's mind, on "the bitter struggle of race with race, the result of man, like all other life, being subject to the stern law of the survival of the fitter." A war was already upon civilization, and it was "easier to suggest means of eliminating the manifestly unfit as factors of race perpetuation, than to advocate acceptable methods of emphasising the fertility of the socially most valuable members of the community." If a military war seemed likely from the vantage point of 1912, for Pearson "the perpetuation of sound stock in the nation is no less important than a 'two-power standard' of the fleet."[55]

Practicing the Galtonian Creed

I have tried to demonstrate how sympathy was engineered, being gradually extended *to,* in Darwin's terms, "the men of all nations and races," and arising *from,* in Pearson's terms, the "wreck of nations" and destruction of "inferior races." And I want to suggest that these were not mere words on the page; that this is not merely an intellectual or rhetorical history. On the contrary, I want to suggest that the continual repetition of these ideas—the conviction that humanity really was as scientists described it—genuinely affected the emotional and moral compasses of the men who propounded such ideas. In drawing a scientific picture of life, Pearson was ultimately also advocating the future of humanity through a life of science. Only through scientific training could the higher plane of intellectually guided and far-seeing sympathy be reached. Through a science of breeding he envisaged a future, in his own nation, of managed selection through which the pain and suffering inherent in nature could be eliminated. And this would ensure national triumph in the struggle with other races. This very method was to be the carrier of the ultimate emotional satisfaction, both at the level of the individual who planned and executed it and of society as a whole. His scientific racial social nationalism would represent for him, were it successful, the high points of art, aesthetics, science, and welfare, the pinnacle of civilized society, and the mastery of his own race. And all of this, he thought, would be experienced by the scientist as an emotional satisfaction, a harmony of taste and cosmos, a higher plane of sympathy that left behind all remnants of sentiment.[56]

For all that Pearson wanted to reduce sympathy to statistics, to approach suffering through reason, and to make policy *academic*, he was nevertheless adamant that where science was cool, scientists were moral beings. Where they made prescriptions for the sympathetic attitude of society, they tried to follow it for themselves. Indeed, only the correct moral character would have the necessary strength and habits of mind to displace the ready sentimentalism of society. The scientific character was put to the test when, in 1906, Pearson's closest colleague and fellow founder of the journal *Biometrika*, Raphael Weldon, died suddenly after an acute bout of double pneumonia.[57] This followed shortly after the death of Galton's sister. The correspondence between Galton and Pearson at this time, and particularly Pearson's enormous obituary of Weldon, illuminate the personal experience of sympathy of those men who professed to understand it scientifically. When Galton's sister died, Pearson sent his "very heartfelt sympathy" to Galton, elaborating: "[O]ne feels very strongly the closeness and the mystery of death; and sympathy—which one is helpless to express—goes out to a friend in like case. I have often thought the only real expression of a feeling like this is given by the hand and eye, and not by the tongue, which is so helpless that we had better go on with the old routine of life, speechless on such points."[58] Galton responded with gratitude, exclaiming, "To what an enormous amount of grief do the tombstones of any churchyard bear witness!"[59] Two months later, Galton was writing to Pearson on the "terrible and disastrous blow" of Weldon's death. The personal was already intermingled with the professional. "Few if any men will feel it more deeply than you who were so intimately associated with him," Galton wrote, adding, "We have lost a loved friend, and Biometry has lost one of its protagonists."[60] Biometry was that science devised by Pearson and Weldon statistically to measure biological differences and abnormalities within and among different races.[61] The burden of ensuring its success was now to lie almost entirely with Pearson. Pearson notes in his memoir of

Galton that "Without Francis Galton's continuous sympathy, aid and counsel, it would have been impossible in that year to continue my work."[62] As Daniel Kevles has pointed out, that work was carried out in an obdurate, polemical fashion. "When it came to biometry, eugenics, and statistics," Kevles tells us, "he was the besieged defender of an emotionally charged faith."[63]

Pearson's obituary of Weldon, who was said to hate biography, began with an apology that nevertheless justified the construction of Weldon's own life from infancy. "If there is to be a constant stream of men," Pearson wrote, "who serve science from love as men in great religious epochs have served the Church, then we must have scientific ideals of character, and these do involve some knowledge of personal life and development." Neither the reference to the affective practice of science, nor the allusion to religion, was accidental. Pearson was Galton's high priest in pursuing the idea of eugenics as a creed to replace or reform the Church, and he saw that further priests would be necessary: "[S]cience," he said, "no less than theology or philosophy, is the field for personal influence, for the creation of enthusiasm, and for the establishment of ideals of self-discipline and self-development." Indeed, Pearson averred that "no man works effectively without a creed of life, that for width of character and healthy development there must ever be a proper balance of the emotional and the intellectual." Pearson's testimony to his friend and colleague elevated the character of the man above the quality of his research: "No man becomes great in science from the mere force of intellect, unguided and unaccompanied by what really amounts to moral force," Pearson wrote. The components of that moral force were explicitly laid out: "Behind the intellectual capacity there is the devotion to truth, the deep sympathy with nature, and the determination to sacrifice all minor matters to one great end."[64]

Given Pearson and Galton's apprehension of "nature" as being something that slowly and painfully extirpated the weak in favor of the strong, the phrase "sympathy with nature" should give us pause. The scientist who had a "deep sympathy with nature," in Pearson's eyes, was the scientist who appreciated the "good" of the destruction of the weak in favor of racial improvement, and went about abetting that process, speeding it up, making it more efficient. This should also give us pause about the notion of "sacrifice." As we have seen, Pearson saw the ultimate good that came of the "wreck of nations." If Galton and Pearson put forward eugenics as a creed, or religion, then they also strove to be true believers, to act, think, and feel in these terms. The loss of Weldon was felt as the loss of a true believer, a prophet of the Utopia they were to fashion together. Such a loss was worthy of lament, unlike the childish piteousness of most other people. Pearson found this in common with Galton, of whom he noted, "[H]e did understand and sympathise with those simple childlike natures which still found comfort, and a crutch for the conduct of life, in the faiths of mankind's infancy. He would endeavour to interpret their conceptions in terms of his own wider aspirations. To those who stood nearer to his own standpoint he made no pretence of reconciling the old with the new—[saying] 'It aids them, but it would be of no service to you and me.'"[65] The moral economy they were forging could still build a rhetorical bridge to the large moral economy of society in which they had to work, but the distance between the two was growing.

Pearson's biographical sketch of Galton is consistent with the latter's theoretical approach to religion. Because most could not be expected to understand the science of eugenics, they would be compelled to follow its edicts as a form of duty, through rituals of marriage and breeding that would be put in place. Ignorance could be diverted to more universally useful ends. He would employ Mill's definition of the essence of religion: "the direction of the emotions and desires towards an ideal object, recognised as rightly paramount over all selfish objects of desire." Galton foresaw a merger of the science of evolution and eugenics with the principal institution governing the laws, customs, and mores of marriage.[66] Pearson summed up Galton's views by boiling off the spiritual content of religion to leave its social function. Whether "we regard it as a supernatural revelation or not," Pearson wrote, "we can agree that one of its [religion's] chief functions is to curb selfishness in the individual, to inculcate altruism, and by restraining human passions to help the stabilisation of society." Religion had been both emotional and moral compass, as well as an emotional restraint. It was the "guardian of tribal custom in regard to marriage, birth and death." In all these things, therefore, religion "concerned itself with matters" that evolutionists had since come to see as having "a bearing on human evolution" through the "laws of natural selection and heredity." No longer, according to this view, could "the scientific doctrine of evolution and the moral conduct of man as inspired by religious belief" be placed "in separate idea-tight compartments." Pearson saw glimmers of recognition of the new science among the "more thoughtful clergy," but it was clear that the new priests would replace them. The evolutionary "aspect of religion" "formed the most hopeful field for co-operation between the old supernaturalism and the new scientific knowledge. It is from this conception that Galton, as an agnostic, starts to bring religion into touch as a living force with our belief in human evolution."[67] Galton's ease with the "childlike natures" of the faithful could thus be understood, and by this means Galton explained to Pearson his "sympathy in expression and action which might not unreasonably appear irreconcilable with his own faith."[68] The masses could be brought to believe in a higher power if necessary, if it served eugenic ends. In this sense, Galton and Pearson were consciously carving out roles for themselves as civil guardians, philosopher kings in the Platonic sense.

It is noteworthy that after Karl Pearson's death, his son Egon wrote a memoir of his father in *Biometrika*, quoting from Karl's memoir of Weldon some thirty years previously. He observed of his father that the "moral force, this 'creed of life'" that Pearson had noted of Weldon "was developed in Pearson to a marked degree." In a manner of which Galton would surely have approved, Egon Pearson asserted that "In the life of Karl Pearson we may trace [. . .] the development of a new faith and of the blending of this faith with his outlook on science."[69] Indeed, Egon himself was the product of a eugenic marriage. Pearson had founded the Men and Women's Club in 1885 for the purposes, so Kevles avers, of teaching himself about women. He had courted his wife to-be, Maria Sharpe, "by abstract discourse and correspondence on socialism, women, and sex," aiming to convince her, at least theoretically, that sex might in the future be recreational rather than procreational. The thirty-something Sharpe, independent and feminist, suffered what Kevles describes as a "nervous breakdown" when Pearson proposed, but after six months of "melancholia" she and Pearson tied the knot. Pearson was living out his deeply held theories, cementing his ideas in and through practice.[70] He and Galton were engaged in an emotive process in which they made theories about the science of morality and emotion based on prior political orientations and personal assumptions, thereafter feeding back their own scientific discourse into their personal and interpersonal practices. They were striving to match their own feelings and expressions with the standards they had set for a highly evolved civilized society, authenticating their own experiences by their theories, and their theories through their own experiences. Karl Pearson and Francis Galton doubtless thought themselves exemplary moral and sympathetic men, working tirelessly for the good of their society. Insofar as we can locate them within a moral economy of science in which sympathy involved a dynamic process of theoretical and political exposition, scientific practices, emotional expressions, and interpersonal relationships, we can probably attest that, in their own terms, they were indeed exemplarily sympathetic. But crucial to this observation is the acknowledgment that this sympathy was not, in this context, sympathy as it was known, either before or since.

Heretics in the Camp

Given the force of expression of Galton, and especially Pearson, and their conviction that their understanding of sympathy, philanthropy, and humanity was a true expression of that which Darwin had prescribed, why did a program of eugenic policies fail to emerge in Britain? Pearson and Galton were not working alone, and not in a vacuum. Eugenics was developed in the public eye, constantly subjected to public opinion, and often with the "help" (unwanted input) of interested laymen. This often served to blur the message Galton wished to promote. Despite the dominance of Galton's view and the volume of Pearson's advocacy, other Darwinists continued to pedal a softer line, more in keeping perhaps with Darwin's own. Nobody was immune to the social implications of Darwinism. Darwin himself saw it from the first. But there was much great generosity of spirit in the definition of the "weak" among some thinkers, who saw in the expanded reach of public health, armed with new expertise, the transformation of the apparently weak into new embodiments of the strong. It is vitally important to recognize that the future of sympathy was heavily contested, and that different scientists decided to live Darwinism in different ways.

Both T. H. Huxley and Alfred Russel Wallace sharply diverged from the Spencerian position, arguing that natural selection no longer functioned within civilized society, and that to make moral exhortations about the preservation of the weak was to misunderstand fundamentally what those people who appeared to be weak really were. Huxley pondered on the wisdom of the aphorism that "dirt is riches in the wrong place." For, he said, the "benevolence and open-handed generosity which adorn a rich man, may make a pauper of a poor one; the energy and courage to which the successful soldier owes his rise, the cool and daring subtlety to which the great financier owes his fortune, may very easily, under favourable conditions, lead their possessors to the gallows, or to the hulks." He therefore took aim at those men who

> are accustomed to contemplate the active or passive extirpation of the weak, the unfortunate, and the superfluous; who justify that conduct on the ground that it has the sanction of the cosmic process, and is the only way of ensuring the progress of the race; who, if they are consistent, must rank medicine among the black arts and count the physician a mischievous preserver of the unfit; [. . .] whose whole lives, therefore, are an education in the noble art of suppressing natural affection and sympathy.

He concluded that such men "are not likely to have any large stock of these commodities [of affection and sympathy] left."[71] Wallace was even firmer in his declaration of human exceptionalism in civilized society, in which the callous vagaries of natural selection were no longer in play: "[O]ur legislation," he said, "should be for the saving of human life and not saving certain persons while permitting the destruction of others."[72]

Though few establishment scientists went along with Wallace in the specifics of his politics and his increasingly esoteric approach to evolutionary biology, the establishment had firmly entrenched this principle of saving life. John Simon, who had pronounced on the medical advantages of vivisection, and who had overseen the public health revolution of compulsory vaccination, also raised his voice against eugenic ideas. The influence of such a voice should not be underestimated. Simon's was the voice of practical experience. He could not be written off as overcome by sentimentalism, or as insufficiently grounded in scientific theory or medical practice. He understood, perhaps better than anyone ever had understood, the epidemiology of smallpox, and he understood that the propensity to catch it had nothing to do with heritable weakness. In retirement, he offered forth his collected thoughts in the volume *English Sanitary Institutions,* appending to it an essay he had written a few years previously "On the Ethical Relations of Early Man." The two texts join up the dots of Simon's theoretical understanding of society in evolutionary terms with his policies on public health and experimental medicine. Simon was a Darwinist in the mode of Darwin himself in *Descent.* Common action, which Simon chose to call "altruism," worked, in his view, for the welfare of the community. There was no question of leaving that community, comprised of collective individuals, to fend for itself.[73] Dear bought and long experience had shown that "the success of any given race greatly depends on the degree in which the individuals of the race combine for their struggle, and are helpful in it each to the other." The natural history of society hinged on the deliberation of how to reconcile individual effort (Egotism) and common duty (Altruism) for the common good. The outcome of this deliberation defined morality. Naturally, as social conditions changed, so the moral yardstick would change concurrently.[74]

Simon opens his account with a congratulatory note to all the agencies that assisted the "great system of Preventative Medicine," from those with moral influence, to those with educative or economic influence, to those "judicious public or private organisation[s]" that offer "kindly succour and sympathy to the otherwise helpless members of the community."[75] Clearly such a statement pegged Simon as an opponent of Galtonian ideas, but Simon would go on to explain why a eugenic religion could not work. He imagined a primitive human tribe that suddenly finds itself in "extreme difficulties of struggle," and how it might consider whether to kill or to cut off the food supply of "ineffective lives." But any tribe that "exercised prerogatives of life and death could not exempt itself from the common conditions of morality, but must at least by degrees learn standards of right and wrong for its estimate of difficulties and its application of expedients; and to adjudicate between life and life, between expedient and expedient, would soon lead human thought into the depth of morals." Here the question and concept of kin might trump the expediencies of a calculated cull. The quandary, for Simon, was familiar:

> When tribes or families had begun to consider under what pressure of exterior circumstances they would be ready to leave their weaker kinsfolk to starve, or would abandon first dictates of Nature in the relations of sex to sex, and of sexes to progeny, the moral questions before them were essentially of like kind with the questions which engage modern thought; and it may safely be assumed that, as soon as such questions arose, lines of cleavage, such as are now familiar to us, began forthwith to reveal wide distinctions in the moral structure of mankind.

Simon reclaimed the "noble" nature of humanity that Darwin had divined, noting that "natures of the nobler type would practise and proclaim the altruism which identifies the welfare of others with its own," and contrasted such nobility with the "rude egotism" that serves its own appetites by any expedient measure. Where a tribe was lucky enough to be able to boast individuals with altruistic instincts, it would find these individuals pleading with tribal councils "against the ruthless putting away of old and young." The result: "tribes of improving quality would more and more think it shameful to draw strength from the life-blood of the weak, or to thrive by cruel and obscene practices against Nature." Following Darwin, this "shame" would affect intertribe relations, until a "future could be foretold when many peoples would have as it were but one conscience, and would cease from inflicting cruelties on each other."[76]

By such an interpretation, Galton and Pearson were savage throwbacks, rejecting the very qualities that had brought them about, sinking the future of humanity into an equation of bald expediency. Simon was convinced that an "educated spectator" of the "toils and sufferings of the very poor" would understand the "*biological* meaning of what he sees." It was nothing more than "the still-continuing aboriginal struggle of mankind for existence." True enough, however beneficent the ruler, the laws of evolution erected "iron walls" that could not be breached. Success would not—could not—be universal.[77] But all the same, the "constantly increasing care of the community at large for the welfare of its individual parts is an eminently characteristic and influential fact in our present stage of civilisation." It represented the continuous development of "the philanthropic spirit" of humanity. Much of this was superficial, "comparatively crude and impulsive," but even this was significant of future good to come. But for the "more thoughtful classes," a deeper realization had set in. Civilization was tenuous, subject to collapse, a mere veneer, unless and until "the mass around which they are cast be pervaded from rank to rank by the kindliness of man to man."[78] So long as eminent men of science continued with such a line of thinking, the racial warfare drum beaten by Pearson and his kind would sound hollow.

If there was to be a new creed, it was to be the loyalty of man to his own kind. In the future, "so far forward as man's moral outlook can reach, they who shall be in the front will more and more have to count it sin and shame for themselves, if their souls fail of answering to that high appeal, and they strive not with all their strength to fulfil the claims of that allegiance." Though Simon dealt in terms of egotism and altruism, his final words on the matter struck a more familiar note. Picturing in his mind the "stone-breaking cave-dweller whom our present generations call their ancestor," Simon marveled at the distance modern science had traveled. But he marveled equally at the advancement in "conceptions of social duty." His final word of gratitude was in equal measure to the "great Interpreters of Nature" and to the men who "have been the organisers of help and hope for their kind, and have made Human Sympathy a power in politics."[79]

Simon found echoers from disparate sources, which, despite the swelling tide of the eugenics movement, acted as bulwarks against that movement becoming the religion of Galton's dreams. The Scottish naturalist John Arthur Thomson occasionally marked the "appalling glimpses" of the struggle for existence, but admired the civilized insistence on "keeping the unfit alive," borne in part "through genuine sympathy" and in part "from a desire to avoid unpleasantness."[80] Doubtless, the evolution of civilization had tended toward the overthrow of "the yoke of natural selection," but the dilemma this brought about could only be met in one way. "It is impossible," Thomson said, "to return to a natural selection regime, and yet we have not been able to put an equally effective social selection into operation." The solution offered by the eugenicists, in his view, was to "restrict our kindness" lest the future call that kindness "cruelty." Did not the prevention of the "elimination of weaklings and wasters" only cause a "drag on the race?" Perhaps, but there was no getting away from the established fact, so far as Thomson was concerned, that "without a great change in social sentiment, it is in civilised communities quite impossible not to try to save those to whom Nature would show no mercy." Maybe this hurt us in the long run. Maybe "we are often cruel in our charity." Yet as Darwin had argued, "we cannot altogether help it," and nor should we.[81] "Perhaps the time may come when the noblest social sentiment and a maturer science will agree that this bud and that should not be allowed to open," Thomson argued, "but the time is not yet." After all, who was of sufficient intellect, or omniscience, to see, say or do such things? "It is one thing to discourage in every feasible way—compatible with rational and social sentiment—*the breeding of weaklings by weaklings*; it is another thing to look a fellow creature in the eyes and say, 'You must die.'"[82] Though Thomson supported the general thrust of eugenics, construed positively, of which there was "no nobler enthusiasm," he knew that "social sentiment" would "not permit social surgery."[83]

History, at least British history, has so far borne out Thomson's claim. Eugenic ideas became increasingly popular in the first part of the twentieth century, but Galton's and Pearson's dreamed-for religion, in which citizens dutifully practiced statistical discretion in race betterment, never came to pass. In the process of the expansion of ideas, somewhere along the line the foundational importance of a new sympathy got lost. Eugenics advocates were more likely to express fear of degeneration, rather than optimism for controlled selection and a better society. The vision of society put forward by John Simon, in which the modern struggle for community survival was not essentially of a difference in kind to the primitive struggle for survival, and in which the ever-expanding compass of human sympathy continued to work for the betterment of civilization, prevailed. In a sense, the first part of the Darwinian forecast, of a sympathy for other communities, other nations, and even other animals, looked to have come to pass by the end of the nineteenth century. The outstanding concerns about degeneration and physical deterioration, brought bluntly to the attention of the public by the condition of soldiers at the time of the Boer War (1899–1902) and by Charles Booth's *Life and Labour of the People in London*, were not met with widespread calls for eugenic policies to be enacted.[84] Rather, they were met with calls for state assistance, better diet and sanitation, better education (both formal and moral), restrictions on the sale of alcohol and tobacco, better medical science, and better working conditions.[85] The impetus, insofar as Darwinian sympathy was becoming enshrined at the level of the state, was in the direction of universal sympathy, where the weak were not to be left behind. It seems that Darwin's initial fears of an inevitable decline wrought by the preservation of the weak were not shared by the majority. Or, if they were, the majority would perhaps have agreed with Darwin's other assertion, that to let the weak go to the wall went against that which is most noble in humanity.

Conclusion
Scientism and Practice

During the research for this book, I was asked on several occasions whether there was really anything left to say about vivisection, vaccination, and eugenics. "Hasn't all this been done? Do those historical topics really need another rehash?" New historical approaches are always opening up, and in revisiting large bodies of scholarship and testing it with new tools, new insights emerge, new analyses are formed, old assumptions are cast aside. The guardians of historiographical orthodoxies inevitably object to revision. They see the work of a generation trampled underfoot. But the bastions of orthodoxies were themselves once perceived by others as revisionists, at which time they protested that they were standing on the shoulders of giants, seeing farther by dint of the work already done, to which they were adding their own new ways of seeing, researching, and asking questions. The history of emotions affords us a whole raft of new insights that behoove us to revisit old questions. Until now, for example, there has been very little analysis of the emotional engagement of vivisectionists and their opponents, or of the institutions of vaccine administration and their opponents. The emotional basis of eugenic ideas has been completely unexplored.

It might be objected that, especially with respect to vivisection, there has been plenty of scholarship on sentimentality, and on the callousness of the physiologists. But this work has tended to take such historical emotional labels at face value. There has been no sophisticated historical analysis of emotional categories in this context. Likewise, the excellent work on the vaccination question has tended to prioritize the anti-vaccination movement and the intrusion of the state into the private lives of citizens. The contemporary moral justifications for this intrusion tend to be overlooked, as do the competing visions of sympathy that activated the debate. The historiography of eugenics is enormous, and constantly expanding, but nobody has ever analyzed the emotional component that was the essential foundation stone for the formation of eugenic ideas. The oversight, I suspect, is not based on the obscurity of my argument. A certain, perhaps corrupted, Darwinian vision of sympathy is endlessly repeated, especially in the works of Pearson. But this sympathy was so unfamiliar, so at odds with what *we* know to be sympathetic, that it is easier to pass by it than to try to explain it.

It is in the endeavor to explain such an unfamiliar notion that this book becomes more than the sum of its parts. It is not a book about physiology, then vaccination, then eugenics, but a book about what these diverse areas of medical and scientific practice and thought have in common as justifying principles and as functioning affective practices. They are each spheres of scientific practice that say something similar about what it meant to be and how it felt to be a practicing scientist at the end of the nineteenth century. Though there was not uniformity of opinion among all scientists, medical researchers, and science popularizers in this period, what does emerge is a clear moral economy of science that informs and delimits both intellectual debate and the possibilities for emotional expression and affective practice. Sympathy was transformed by Darwin's *Descent of Man,* and for those who subscribed to its principal doctrines, everything changed. Sympathetic reactions had to be put through measured judgment. The instinctive or habitual sympathies of others had to be critiqued as false, egocentric, and ultimately of limited or no benefit to society. Their own emotions, with relation to religious compassion, to the sight of blood, to the commission of pain, to invasions of privacy, and to liberty, had to be checked, changed, redirected. The process was not always smooth. Emotive failure lurked at every turn. Yet the community of scientists was sufficiently large to provide continual reinforcement that this new emotional regime, this new moral economy, was on a right footing, was morally and emotionally justified. Everyday practice of the "science of sympathy" led to a widespread embodiment of the feeling of "scientific sympathy," and vice versa.

That this brand of sympathy might seem so strange to us, and in some cases seems to justify actions that we might find not merely immoral but bordering on the criminal, should give us pause. Science was—is—never neutral. Objectivity, as Lorraine Daston and Peter Galison so brilliantly observed, is an affect. Claims that advert to an unimpeachable distanced observation are always laced with a degree of dogma, however implicit or unconscious. The view is always from *somewhere.* Appeals to science have a tendency to be appeals to incontrovertibility, and are opposed to other forms of knowledge that are pejoratively reduced to mere beliefs, opinions, feelings. The nineteenth century witnessed the first high point of this insistence on the primacy of positive knowledge, a material empiricism that would trump the mystical and the magical. But in its insistence on assuming priority in the order of systems of knowledge, it rejected the reflexivity that it demanded of everything and everyone else. The emerging class of scient*ists*, a nineteenth-century neologism that tried to capture the distinct activities of this diverse group of knowledge chasers, ran in parallel to the emergence, even the conscious construction, of a more or less fundamentalist scient*ism*.

It might be objected that such a label befits some parts of this story better than others. Pearson and Galton were explicit about their science-as-religion purposes, and were busy setting about the construction of grand Platonic lies to get the population on board. Others, not least Darwin and his disciple Romanes, were more tentative. Their tentativeness was partly borne on the wings of fear, that society was not ready for such a seismic shift in first principles and that being too pushy might just be at the expense of science. It was also partly borne on the wings of doubt, for men like Romanes in particular. He took Pascal's wager, returning to the church in his troubled later years when a brain tumor signaled that the end was nigh. He resisted Huxley's antitheistic campaigning, explaining that "the old landmarks in the territory of religious belief are now being hurried away at a rate which scarcely seems to call for any organized effort to ensure their more rapid destruction."[1] He was right, but the note of despair is also palpable. To be clear then, I am not arguing that there was not a range of opinions, feelings, and politics in play among this first generation of Darwinists. But taken as a whole, when compelled to act, be it procedurally or politically, there was a public coherency, a dogmatic scientism.

Central to this moral economy was a loosely shared feeling that the actions of scientists, informed by a new natural-historical cosmology that Darwin had pioneered, were beneficent. Scientists were at the apex of humanity, and that humanity was confirmed, enacted, embodied in the practice of their respective disciplines. Whatever individual doubts they might have maintained, about God, society, civilization, morality, they nevertheless actively appropriated the role of do-gooder from the realms of religion, charity, and sentimentality. The new recipe knowledge gained by evolutionary theory and the culture of experiment made new forms of ethical practice not only likely, but necessary. The *doing* of good, rather than merely intellectualizing or philosophizing about it, is of central importance.

The first generation of Darwinists were not merely social commentators or critics. They affirmed their world views through activities—cutting, culturing, calculating—affectively practicing what they preached to be a higher form of sympathy, which, when enacted, was a higher form of morality. This classic emotive process, of striving to match feelings to feeling rules through *utterances*, in the broadest meaning of the term, was undertaken collectively. Individuals looked to each other for positive reinforcement, finding and testing the limits of this emerging moral economy. What is distinctive and important about this moral economy is that it immediately assumed a dominant position in society as a whole. The gentlemen, public men, elite men, who comprised it presumed a natural—literally—authority to disseminate their views to the lay public, and to displace those who dissented from it. Their public status ought to have set them apart as social conservatives, traditionalists, and reliable bastions of public opinion. Their science suddenly made them into ardent radicals, but in dangerous positions of influence. Yet this radical position was not only presented as natural, but also as neutral. There were no appeals to mysterious or otherworldly entities to justify the authority of this moral economy. Everything was to be testable and observable. But the capacity to test and observe was vastly overestimated. The ability to reflect on starting assumptions and intellectual blind spots was often lacking, coupled with a tendency unwittingly to manipulate experimental data to produce expected results. No wonder the opponents of science were terrified of the consequences to civilization.

This community of scientists were extremely well equipped to critique the outgoing model of morality based on religious belief. They rarely entertained the holes in their own cosmology, but I suspect that such reflexivity was extremely difficult if not actually impossible. Part of the new faith was that the missing pieces—of evidence, of knowledge, of understanding—would be revealed in the course of time, if only the course of experimentation and pursuit could be followed without let or hindrance. It sounded so much more reassuring than the vain hopes for divine revelation or illumination. This was an empirical creed, and revelation would be testable, provable, or else refutable.

Of course, we know now that much of Darwin's work has not endured particularly well. He was, strictly speaking, wrong about many things. A central idea has been extracted, developed, tested, and proven, but that idea was just one strand in a haystack of jumbled hypotheses in Darwin's own lifetime. He was a latter-day Lamarckian, unconvinced in the all-powerfulness of his own theory of natural selection. Among even his most dedicated allies, natural selection was not the be-all and end-all of Darwinism. But the implicit hope of evolutionary science, with its bedfellows physiology, toxicology and immunology, and eugenics, seemed so much more immediate, worldly, tangible than anything that had gone before. Early gains in understanding and technique, along with medical breakthroughs, inspired great confidence—hubris—in these new aficionados of the science of sympathy. Where reflection did take place, as with Romanes, crisis seemed to be the only likely outcome. Darwin had persuasively demonstrated the lack of need for God, the implausibility of intelligent design. If one rejected religion and then doubted Darwin, what was left? Hence Romanes's flip-flopping between the two. He did not know how to feel, or what to do. His emotive failure was extreme. The rest went about their business with, largely, unswerving conviction. In the name of neutrality, objectivity, and empiricism, a scientistic ideology of sympathy gained ground rapidly.

The space carved out by this new ideology was populated by new selves. Each of the principal men of science covered in this book underwent a more-or-less conscious re-evaluation of what it meant for them, personally, to be human. They had been through individual processes of self-construction, mediated by the emerging collective around them, in which they found justifications for feeling positively about and through what they did. They also justified what they did through asserting and internalizing the claim that their actions were *good*. This was not merely circular logic, but an emotive process: a dynamic relationship between thinking and feeling, feeling and doing, motion and emotion. New prescriptions for emotional expression, moral intentions, and moral outcomes were embedded in new practices and institutions that gave enlarged meaning to a whole raft of mundane (though often unpalatable) practices. At the same time, new institutions and new practices were established precisely because of the meanings attached to these practices. As the moral economy took shape, it provided its own expansionist justifications. Within this space, decisions looked obvious, natural, and moral. Outside this space, the expansion, both of institutions and of the numbers of men to populate them, seemed bewilderingly rapid, dangerous, and incomprehensible.

What happened to this scientistic ideology of sympathy? This is perhaps the most alarming part of the story. My argument throughout this book was that a new form of sympathy was the foundational principle for new scientific practices. It justified them, both emotionally and morally. Yet if you read any of the abundant literature on the history of eugenics in the twentieth century, for example, you will not find a single reference to sympathy (even though you will find plenty to Darwin). Similarly, the history of physiology, and its continued practice, might argue for its beneficial ends, but the moral/emotional grounding has largely disappeared. If the public hears about vivisection at all, it is usually from the street purveyors of animal rights, bemoaning the blood on the hands of science. Yet vivisection, or animal experimentation in general, remains a central research technique in physiology, toxicology, immunology, neuroscience, and genetics research. Where research is mentioned in prominent places, ethical misgivings are downplayed in the context of promising results. We—that is, the public—are put off worrying about motivations and intentions because we tend only to be presented with outcomes. Those outcomes have always already justified the means. Prior justification simply is not part of everyday discourse any more.

Science has its own inertia. Institutions, funding, experts, personalities, and so forth, ensure that what has happened is what will happen. Procedure is paradigmatic. Change is incremental. Results are slow in coming. Applications of these results are even slower in coming. But these things are ingrained only after a successful period of establishment and institutionalization. Practices have to be practiced, honed, developed. Technology requires innovation before it can rely on tinkering. The period covered here is just such a period of profound expansion, invention, and establishment. The science of sympathy, encompassing everything from the theory of the evolution of moral societies to the embodied and expressed feelings of individual scientists, was necessary for this preliminary work, this establishment work, to be carried out. In the teeth of a vigorous opposition, picking out the immorality, the callousness, and the cruelty of the new science, a more than rhetorical claim to a higher form of feeling was critical. Scientists not only had to talk a good game. They had to believe it themselves. They maintained this for a generation, after which time reality had shifted. Justifications were not so necessary. Feet were under the table. Some important discoveries—not least in the field of immunology—had made good on some of the promises. Meanwhile, a new generation of scientists picked up the reins of practice without needing to strive to buy into the ideology, or the feeling rules, and without needing to fight so fiercely for their right to work, to exist.

The prevailing form of sympathy, that defined by charity, religious sentiment, and an aesthetic sense of disgust at the sight of blood and the commission of pain, quietly reassumed a cultural dominance in the lay world as science increasingly retreated behind the closed doors of professionalism. The First World War, with its all too palpable horrors of physical and mental suffering, made a loud call on the sympathy of the societies involved, and the sympathy called forth had little to do with that which animated the physiologists or the eugenicists. Here was bald human pain, crying out for succor.

It was a dramatic reversion to a longer-lasting creed of sympathy as social glue, of the Golden Rule of doing good to others, as one would wish done to oneself. The war was also a stark reminder that civilization was not at such a high pitch as the late Victorians had hoped or presumed. Galton's dream of a scientific priesthood, and of a eugenic religion, never came to pass. But the science of sympathy existed for long enough to ensure the lasting legacy of these men of science. Their work found traction, and has been enormously influential, but in a process of divorce from public life.

In this case, an emotion (or emotional mechanism) was theoretically recast, disrupting commonly held assumptions about its origins, meanings, and implications, and employed to give moral justification to new kinds of practice done by new kinds of people. The striving of these people to feel and act anew, to fulfil the requirements of new prescriptions about emotional expression and moral outcomes, led to dramatic changes in scientific practices that had direct consequences in the lives of those beyond the moral economy of science. Historians look for causes and effects, attempting to explain change. Historians of emotion are beginning to assert that emotions themselves are causal. From the evidence presented here, my challenging conclusion is that the rapid expansion of physiology, the determined adherence to a policy of enforced public health, and the founding of the principles of eugenics, could not have come to pass without a major (if temporary) disruption of assumptions about the origins, meanings, and implications of sympathy, embodied by a large group of influential men. What I hope to have demonstrated is the power of emotions to make history.

Notes

Chapter 1. Emotions, Morals, Practices

1. Charles Darwin, *The Descent of Man, and Selection in Relation to Sex* (London: John Murray, 1871). All quotations from *Descent of Man* in this book are from the second edition of 1879 (London: Penguin, 2004). The principal passages on sympathy are 119–72.

2. For the evolutionary hierarchy of beings, including differentiation among human beings, see Rob Boddice, "The Manly Mind? Re-visiting the Victorian 'Sex in Brain' Debate," *Gender and History* 23 (2011): 321–40. For examples of Darwin's racial and intellectual ordering of man, see Darwin, *Descent of Man,* 86, 114–15, 116, 118, 119, 125, 133.

3. Darwin, *Descent of Man,* 133.

4. The idea of the drowning stranger was mooted by Herbert Spencer, *The Study of Sociology* (New York: D. Appleton, 1874), 360–61. See Thomas Dixon, *The Invention of Altruism: Making Moral Meanings in Victorian Britain* (Oxford, U.K.: Oxford University Press, 2008). Dixon's brilliant analysis of the word history of altruism nevertheless underplays the extent to which sympathy remained the overwhelmingly dominant mode of expressing an emotional foundation of doing good to others. The emergence of altruism refined existing and changing notions of sympathy, but remained an epiphenomenon of sympathy.

5. The construction belongs to Spencer, not Darwin, in his *Principles of Biology* (London: Williams & Norgate, 1864), I, 444.

6. Charles Darwin, *The Origin of Species by Means of Natural Selection or The Preservation of Favoured Races in the Struggle for Life* (1859; London: Penguin, 1985).

7. Prinz himself offers a comprehensive survey of this facet of the history of philosophy in his *The Emotional Construction of Morals* (Oxford, U.K.: Oxford University Press, 2007). See also Daniel M. Gross, *The Secret History of Emotion: From Aristotle's Rhetoric to Modern Brain Science* (Chicago: University of Chicago Press, 2006), esp. chapters 4 and 5.

8. M. J. D. Roberts, *Making English Morals: Voluntary Association and Moral Reform in England, 1787–1886* (Cambridge, U.K.: Cambridge University Press, 2004); Norbert

Elias, *The Civilizing Process: Sociogenetic and Psychogenetic Investigations,* trans. Edmund Jephcott (1939; new ed., Oxford, U.K.: Blackwell, 2000).

9. Thomas Haskell, "Capitalism and the Origins of the Humanitarian Sensibility," part 1, *American Historical Review* 90 (1985): 339–61.

10. Ibid., 352.

11. Ibid., 356.

12. Ibid., 357.

13. Ibid., 358.

14. Ibid.

15. The classic text here remains G. J. Barker Benfield, *The Culture of Sensibility: Sex and Society in Eighteenth-Century Britain* (Chicago: University of Chicago Press, 1992). A number of works specifically on eighteenth-century sympathy have appeared in recent years: Michael Frazer, *The Enlightenment of Sympathy: Justice and the Moral Sentiments in the Eighteenth Century and Today* (Oxford, U.K.: Oxford University Press, 2012); Ildiko Csengei, *Sympathy, Sensibility and the Literature of Feeling in the Eighteenth Century* (Houndmills, U.K.: Palgrave, 2012); Paul Goring, *The Rhetoric of Sensibility in Eighteenth-century Culture* (Cambridge, U.K.: Cambridge University Press, 2004).

16. Two recent works have made strides in this direction for the nineteenth century in particular: Javier Moscoso, *Pain: A Cultural History* (Houndmills, U.K.: Palgrave, 2012), chapters 3 and 4; Joanna Bourke, *The Story of Pain: From Prayer to Painkillers* (Oxford, U.K.: Oxford University Press, 2014), chapter 8.

17. Haskell, "Capitalism," 358.

18. Adam Smith, *The Theory of Moral Sentiments* (1759; London: Penguin, 2009), 31.

19. Smith, *Theory of Moral Sentiments,* 27.

20. This is in contradistinction to the "moral economies" employed by Robert E. Kohler or Bruno Strasser, borrowings in turn from E. P. Thompson, which "regulate authority relations and access to the means of production and rewards of achievement." See Robert E. Kohler, *Lords of the Fly:* Drosophila *Genetics and the Experimental Life* (Chicago: University of Chicago Press, 1994), 5; Bruno Strasser, "The Experimenter's Museum: GenBank, Natural History, and the Moral Economies of Biomedicine," *Isis* 102 (2011): 60–96. Daston's usage is specific to "The Moral Economy of Science," *Osiris* 2, no. 10 (1995): 2–24. The concept is implicit throughout her study, with Peter Galison, of objectivity, but replaced with the notion of "epistemic virtue." See *Objectivity* (New York: Zone Books, 2007), esp. chapter 4.

21. Daston, "Moral Economy," 4.The plurality of usages for "moral economy" has caused some confusion, not least for economists, but there is sufficient definitional rigor in Daston's definition to ward off any potential confusion. The conceptual usefulness of the term—according to a variety of definitions—seems only to be growing. See the special issue of the *Journal of Global Ethics* 11 (2015) on "Moral Economy: New Perspectives," and in particular the synoptic contribution of Norbert Götz, "'Moral economy': its conceptual history and analytical prospects," 147–62.

22. Barbara H. Rosenwein, "Worrying about Emotions in History," *American Historical Review* 107 (2002): 821–45; Barbara H. Rosenwein, *Emotional Communities in the Early Middle Ages* (Ithaca, N.Y.: Cornell University Press, 2006).

23. Rosenwein, "Worrying," 842.

24. William Reddy, *The Navigation of Feeling: A Framework for the History of Emotions* (Cambridge, U.K.: Cambridge University Press, 2001), esp. chapter 3.

25. Barbara H. Rosenwein, "Problems and Methods in the History of Emotions," *Passions in Context* 1 (2010): 1–32 at 22–23; Jan Plamper, "The History of Emotions: An Interview with William Reddy, Barbara Rosenwein, and Peter Stearns," *History and Theory* 49 (2010): 237–65 at 255–56.

26. Plamper, "History of Emotions," 243.

27. Ibid.

28. Daston, "Moral Economy," 4–5.

29. Paul White, Introduction to "Focus: The Emotional Economy of Science," *Isis* 100 (2009): 792–97 at 793. This also responds to a call made by Otniel Dror, Bettina Hitzer, Anja Laukötter, and Pilar León-Sanz in "History of Science and the Emotions: Perspectives and Challenges—An Introduction," *Osiris* 31 (2016). My thanks to Bettina Hitzer and Otniel Dror for allowing me to read this in advance of publication.

30. This in part resembles Rosenwein's nonconcentric and intersecting circles of emotional communities, but with a more refined definition of the working of power dynamics, both in the community and in the individual itself, at the level of experience. See Rosenwein, *Emotional Communities*, 24.

31. Alfred Russel Wallace, *My Life* (New York: Dodd, Mead, 1905), II, 368. My thanks to Ahren Lester for this reference.

32. William Reddy, "Against Constructionism: The Historical Ethnography of Emotions," *Current Anthropology* 38 (1997): 327–51; Reddy, *Navigation of Feeling*.

33. See also Arlie Russell Hochschild, "Emotion Work, Feeling Rules, and Social Structure," *American Journal of Sociology* 85 (1979): 551–75.

34. I introduced this term in "The Affective Turn: Historicising the Emotions," in *Psychology and History: Interdisciplinary Explorations*, ed. Cristian Tileagă and Jovan Byford (Cambridge, U.K.: Cambridge University Press, 2014).

35. For an account of the analysis of scientific personae, see Daston and Galison, *Objectivity*, chapter 4.

Chapter 2. Sympathy for a Devil's Chaplain

1. Darwin, *Origin of Species*, 135.

2. Darwin's account of the evolution of sympathy and morality can be found in *Descent of Man*, 119–72.

3. The most emphatic purveyor of this naysaying is Richard Dawkins. See the most recent edition of *The Selfish Gene* (Oxford, U.K.: Oxford University Press, 2006), 8–11. Recent iterations of the evolutionary importance of social connections have stressed their importance for the survival of the individual, not for the survival of the society per se. See, for example, Naomi I. Eisenberger, "The Neural Basis of Social Pain: Findings and Implications," *Social Pain: Neuropsychological Implications of Loss and Exclusion*, ed. Geoff MacDonald and Lauri A. Jensen-Campbell (Washington, D.C.: American Psychological Association, 2011): 53–78 at 53–54.

4. See, for example, Darwin, *Descent of Man,* 83, 130.

5. Susan Lanzoni, "Sympathy in Mind (1876–1900)," *Journal of the History of Ideas* 70 (2009): 265–87 at 272–73.

6. Smith, *Theory of Moral Sentiments.*

7. Darwin, *Descent of Man,* 122.

8. Ibid., 129.

9. Ibid., 137n, 151, 157.

10. To a large extent Darwin resolves the apparent contradictions between sympathy and selfishness in Smith's oeuvre. Darwin, *Descent of Man,* 130, 145–46, 156. See also Dixon, *Invention of Altruism,* 138, 151. Robert J. Richards has detailed the point at which Darwin diverged from Smith and Bain over pleasure and pain in his *Darwin and the Emergence of Evolutionary Theories of Mind and Behavior* (Chicago: University of Chicago Press, 1987), 209–10.

11. Darwin, *Descent of Man,* 130.

12. Ibid., 132.

13. Ibid., 130.

14. Ibid., 133.

15. Ibid., 134.

16. Ibid., 147.

17. Ibid., 147, 149.

18. Francis Galton, *English Men of Science: Their Nature and Nurture* (London: Macmillan, 1874), 260. Darwin to Hooker, July 13 [1856], Darwin Correspondence Project, http://www.darwinproject.ac.uk/letter/entry-1924.

19. R. A. Fisher, "Some Hopes of A Eugenist," *Eugenics Review* 5 (1913–1914): 309–15, at 309.

20. Charles Darwin, *Autobiographies* (London: Penguin, 2002), 54.

21. Charles Darwin, *The Expression of Emotions in Man and Animals* (London: John Murray, 1872).

22. Darwin, *Descent of Man,* 218. In this passage, Darwin is mainly concerned with understanding the functional cause of the flow of tears, particularly with reference to the pains or joys of others. See also Thomas Dixon, *Weeping Britannia: Portrait of a Nation in Tears* (Oxford, U.K.: Oxford University Press, 2015).

23. Charles Bell, *Essays on the Anatomy of Expression in Painting* (London: Longman, Hurst, Rees, and Orme, 1806). The work was later expanded to appeal more directly to students of medicine, anatomy, and philosophy, and retitled accordingly: *Essays on the Anatomy and Philosophy of Expression* (London: John Murray, 1824).

24. Bell, *Essays,* 84–85.

25. Ibid., 88.

26. Ibid., 85.

27. Ibid., 87.

28. Ibid., 101.

29. Darwin, *Expression,* 315–16, 326.

30. Ibid., 334.

31. Ibid., 335.

32. Ibid., 221.

33. Darwin, *Descent of Man,* 118: "No being could experience so complex an emotion until advanced in his intellectual and moral faculties to at least a moderately high level."

34. Daniel M. Gross, "Defending the Humanities with Charles Darwin's *The Expression of the Emotions in Man and Animals* (1872)," *Critical Inquiry* 37 (2010): 34–59 at 49. The misapprehension of Darwin's overall approach to the changeability of emotions, in *Expression* and *Descent of Man,* is evidenced in the following: Paul Ekman and Wallace Friesen, "Constants across Cultures in the Face and Emotion," *Journal of Personality and Social Psychology* 17 (1971): 124–29; Paul Ekman and Wallace Friesen, *Pictures of Facial Affect* (Palo Alto, Calif.: Consulting Psychologists Press, 1976); Paul Ekman, *Emotions Revealed: Recognizing Faces and Feelings to Improve Communication and Emotional Life* (New York: Times Books, 2003). Ekman also wrote the introduction to a recent edition of Charles Darwin's *The Expression of Emotions in Man and Animals* (Oxford, U.K.: Oxford University Press, 2002), taking the opportunity to heavily annotate Darwin's text to suit his own purposes.

35. There have been numerous critical appraisals of Duchenne's work and Darwin's appropriation of it, as well as the legacy of this kind of analysis in contemporary emotions research. See, for example, Danny Rees, "Down in the Mouth: Faces of Pain," in *Pain and Emotion in Modern History,* ed. Rob Boddice (Houndmills, U.K.: Palgrave, 2014); Ruth Leys, "How Did Fear Become a Scientific Object and What Kind of Object Is It?" in *Fear Across the Disciplines,* ed. Jan Plamper and Benjamin Lazier (Pittsburgh, Pa.: University of Pittsburgh Press, 2012).

36. Darwin, *Expression,* 180, in ref. to fig. 2, plate 2.

37. Ibid., 182, in ref. to fig. 3, plate 2.

38. See note 34.

39. Darwin to Romanes, 1881, *More Letters of Charles Darwin: A Record of His Work in a Series of Hitherto Unpublished Letters,* eds. F. Darwin and A. C. Seward (London: John Murray, 1903), 213–14.

40. For an overview, see L. S. Jacyna, "The Physiology of Mind, the Unity of Nature, and the Moral Order in Victorian Thought," *British Journal for the History of Science* 14 (1981): 109–32. Romanes was chief among those pushing for a new discipline of comparative psychology in which animal minds would be seen as differing in degree only (not in kind) from human minds. See George John Romanes, *Animal Intelligence,* 3rd ed. (London: Kegan Paul, Trench, 1882); *Mental Evolution in Animals* (London: Kegan Paul, Trench, 1883); *Mental Evolution in Man: Origin of Human Faculty* (London: Kegan Paul, Trench, 1888).

41. Alexander Bain, *Emotions and the Will* (London: John W. Parker and Son, 1859), 3.

42. Ibid.

43. Ibid., 5.

44. Ibid., 14–15.

45. Ibid., 61.

46. Alexander Bain, *The Senses and the Intellect,* 4th ed. (London: Longmans, Green, 1894), 619–22; Bain, *Emotions and the Will,* 220–21.

47. See note 40.

48. This information is extrapolated from the diagrammatical "tree" of mental evolution that Romanes included as the pull-out frontispiece of his Mental Evolution series, which was not completed.

49. C. Lloyd Morgan, *An Introduction to Comparative Psychology* (London: W. Scott, 1894), 59.

50. William James, *The Principles of Psychology,* 2 vols. (1890; London: MacMillan, 1910), II, 449–50.

51. Ibid., 453.

52. Ibid., 454.

53. Otniel Dror, "The Scientific Image of Emotion: Experience and Technologies of Inscription," *Configurations* 7 (1999): 355–401, at 360.

54. See note 35.

55. Quoted in Philip Mercer, *Sympathy and Ethics: A Study of the Relationship between Sympathy and Morality with Special Reference to Hume's Treatise* (Oxford, U.K.: Clarendon Press, 1972), 31.

56. Smith, *Theory of Moral Sentiments,* 45.

57. Ibid., 17.

58. "Sentimental" as pejorative is a departure from its eighteenth-century connotations. See James Chandler, *An Archaeology of Sympathy: The Sentimental Mode in Literature and Cinema* (Chicago: University of Chicago Press, 2013), xiii–xix, 1–5.

59. Michele Cohen, *Fashioning Masculinity: National Identity and Language in the Eighteenth Century* (London and New York: Routledge, 1996).

60. For a review of this literature see Boddice, "The Manly Mind?" See also Katharina Rowold, ed., *Gender and Science: Late Nineteenth-Century Debates on the Female Mind and Body* (Bristol, U.K.: Thoemmes Press, 1996); Carol Dyhouse, "Social Darwinistic Ideas and the Development of Women's Education in England, 1880–1920," *History of Education* 5 (1976): 41–58; Lesley Hall, "Hauling Down the Double Standard: Feminism, Social Purity and Sexual Science in Late Nineteenth-Century Britain," *Gender & History* 16, no. 1 (2004): 36–56.

61. See the extended discussion in the next chapter, but see also Gertrude Himmelfarb, *Poverty and Compassion: The Moral Imagination of the Late Victorians* (New York: Vintage, 1991), for a thorough introduction.

62. Lanzoni, "Sympathy in Mind," 269.

63. Ibid., 270.

64. This is implicit in Caroline Sumpter's claim that Spencer and Huxley and others saw the continuing development of sympathy as "fundamental to moral progress." Caroline Sumpter, "On Suffering and Sympathy: Jude the Obscure, Evolution, and Ethics," *Victorian Studies* 53 (2011): 665–87, at 665.

65. Herbert Spencer, *Principles of Psychology,* 2 vols. (1899; Osnabrück, Germany: Otto Zeller, 1966), I, 583.

66. Benjamin Kidd, *Social Evolution* (London: MacMillan, 1894), 114.

67. See chapter 5.

68. For Osler, see the seminal biography by Harvey Cushing, *The Life of Sir William Osler,* 2 vols. (Oxford, U.K.: Clarendon Press, 1926), and the more recent treatment by Michael Bliss, *William Osler: A Life in Medicine* (Oxford, U.K.: Oxford University Press, 2007).

69. Osler testified before the second Royal Commission on Vivisection in 1907. See U.K. Parliament, Appendix to the Fourth Report of the Commissioners, C. 3955 (1907), 160, 163.

70. Osler, "Aequanimitas," 3–6. For Osler's support of vivisection, see Cushing, *Life*, II, 794–95; Bliss, *William Osler*, 248.

71. Parliament, Appendix to the Fourth Report of the Commissioners [Royal Commission on Vivisection], 158, 160–61, 163.

72. Aptly, the book in question was authored by "Philanthropos," under the title *Physiological Cruelty; or, Fact v. Fancy: An Inquiry into the Vivisection Question* (London: Tinsley, 1883). Romanes's review is in *Nature* 28 (1883): 537–38.

73. E. Romanes, *Life and Letters of George John Romanes* (London: Longmans, Green, 1896), 271.

74. Jim Endersby, *Imperial Nature: Joseph Hooker and the Practices of Victorian Science* (Chicago and London: University of Chicago Press, 2008), 251.

75. *The Times* (London), April 25, 1881.

76. Ibid.

77. Lady Burdon-Sanderson, *Sir John Burdon-Sanderson: A Memoir* (Oxford, U.K.: Clarendon Press, 1911), 157.

78. George John Romanes, "Mental Differences between Men and Women," *Nineteenth Century*, May, 1887; reprinted in *Essays*, ed. C. Lloyd Morgan (London: Longmans, Green, 1897), 119.

79. Daston and Galison, *Objectivity*.

Chapter 3. Common Compassion and the Mad Scientist

1. John Burdon-Sanderson, ed., *Handbook for the Physiological Laboratory*, by E. Klein, John Burdon-Sanderson, Michael Foster, and Thomas Lauder Brunton, 2 vols. (London: J & A Churchill, 1873); the vivisection controversy has a well-established narrative, but the best two works are Richard D. French, *Antivivisection and Medical Science in Victorian Society* (Princeton, N.J.: Princeton University Press, 1975) and *Vivisection in Historical Perspective*, edited by Nicolaas A. Rupke (London: Croom Helm, 1987).

2. See Patrizia Guarnieri, "Moritz Schiff (1823–96): Experimental Physiology and Noble Sentiment in Florence," in Rupke, ed., *Vivisection in Historical Perspective*.

3. Details of Hutton's attitude toward Germany are compiled from Malcolm Woodfield, "Victorian Weekly Reviews and Reviewing After 1860: R.H. Hutton and the *Spectator*," *Yearbook of English Studies* 16 (1986): 74–91, esp. 83; Robert A. Colby, "'How It Strikes A Contemporary': The 'Spectator' as Critic," *Nineteenth-Century Fiction* 11 (1956): 182–206, esp. 195; Gaylord C. Leroy, "Richard Holt Hutton," *PMLA* 56 (1941): 809–40, esp. 814.

4. *Spectator*, 46 (1873) 1643.

5. U.K. Parliament, Report of the Royal Commission on the Practice of Subjecting Live Animals to Experiments for Scientific Purposes, C. 1397 (1876) [hereafter, Royal Commission on Vivisection], 115–16, 144–46.

6. Ibid., 149, 156.

7. Ibid., 82.

8. Ibid., 16–17, 27.

9. Ibid., 43, 47–48.

10. Michael Worboys, "Klein, Edward Emanuel (1844—1925)," *Oxford Dictionary of National Biography*, Oxford University Press, 2004; online edition, January 2008, http://

www.oxforddnb.com/view/article/57359, accessed February 1, 2012. For Klein's involvement in the controversy, see Richards, "Drawing the Life-Blood of Physiology," 44–45; Bruno Atalic and Stella Fatovic-Ferencic, "Emanuel Edward Klein—The Father of British Microbiology and the Case of the Animal Vivisection Controversy of 1875," *Toxicologic Pathology* 37 (2009): 708–13.

11. Royal Commission on Vivisection, 74–75.

12. Ibid., 183–85, 328.

13. Smith, *Theory of Moral Sentiments*, 38.

14. These fears are the subject of Coral Lansbury's *The Old Brown Dog: Women, Workers, and Vivisection in Edwardian England* (Madison: University of Wisconsin Press, 1985) and Susan Lederer's *Subjected to Science: Human Experimentation in America before the Second World War* (Baltimore: Johns Hopkins University Press, 1997).

15. Oscar Wilde, *The Picture of Dorian Gray* (1891; New York: Mondial, 2015), 26.

16. Ibid., 27.

17. Ibid., 107.

18. Ibid., 110.

19. H. G. Wells, *The Island of Dr. Moreau* (1896; London: Heinemann, 1921), 92, 94. For a compelling contextualization of the Victorian mad scientist in literature, see Anne Stiles, *Popular Fiction and Brain Science in the Late Nineteenth Century* (Cambridge, U.K.: Cambridge University Press, 2012), esp. chapter 4.

20. For an excellent account of the stakes of humanity in *Frankenstein*, see Ben Dawson, "Modernity as Anthropolarity: The Human Economy of *Frankenstein*," in *Anthropocentrism: Humans, Animals, Environments*, ed. Rob Boddice (Leiden, Netherlands : Brill, 2011).

21. Rob Boddice, *A History of Attitudes and Behaviours toward Animals in Eighteenth- and Nineteenth-century Britain: Anthropocentrism and the Emergence of Animals* (Lewiston, N.Y.: Mellen, 2009), 109–11.

22. Frances Power Cobbe, *Life of Frances Power Cobbe* (Boston and New York: Houghton, Mifflin, 1904), 633–34.

23. Frances Power Cobbe, "London's Hecatombs," in *Re-Echoes* (London: Williams and Norgate, 1876), 102.

24. Cobbe, "London's Hecatombs," 102.

25. Frances Power Cobbe, *Light in Dark Places* (London: Victoria Street Society, 1883).

26. Frances Power Cobbe, "Darwinism in Morals," in *Darwinism in Morals, and Other Essays* (London: Williams and Norgate, 1872), 13–14, 30.

27. Frances Power Cobbe, "The New Morality," *Zoophilist* 4 (January 1, 1885).

28. Frances Power Cobbe, *Zoophilist* 4 (December 1, 1884).

29. Cobbe, *Life*, 606–7.

30. Frances Power Cobbe, *Zoophilist* 7 (June 1, 1887).

31. Frances Power Cobbe, "The New Benefactor of Humanity," *Zoophilist* 4 (July 1, 1884).

32. The *Times* (London), April 19, 1881.

33. Quoted in an open letter to George H. Simmons, written by Grace D. Davis, Secretary of the Society for the Prevention of Abuse in Animal Experimentation, November 12,

1908. Countway Library of Medicine, Harvard Medical School, Walter B. Cannon archive, H MS c40, Box 28, Folder 333.

34. Royal Commission on Vivisection, Report, x.

35. R. H. Hutton, "Minority Report," Royal Commission on Vivisection, xxii.

36. The *Times* (London), April 30, 1881.

37. Bain, *Emotions and the Will*, 143–44.

38. Ibid., 142–44; 179.

39. The sources here are rich, and build upon a well-developed historiography on the reception of Darwinism and the interplay of science and religion among the first-generation Darwinists. Frank Miller Turner, *Between Science and Religion: The Reaction to Scientific Naturalism in Late Victorian England* (New Haven, Conn. and London: Yale University Press, 1974) and his "The late Victorian conflict of science and religion as an event in nineteenth-century intellectual and cultural history," in *Science and Religion: New Historical Perspectives*, eds. Thomas Dixon, Geoffrey Cantor and Stephen Punfrey (Cambridge, U.K.: Cambridge University Press, 2010), 87–110; Richards, *Darwin and the Emergence of Evolutionary Theories of Mind and Behavior*; Bernard Lightman, *Evolutionary Naturalism in Victorian Britain: The "Darwinians" and their Critics* (Farnham, U.K.: Ashgate Variorum, 2009); Jim Endersby, *Imperial Nature: Joseph Hooker and the Practices of Victorian Science* (Chicago and London: University of Chicago Press, 2008); David Clifford, ed., *Repositioning Victorian Sciences: Shifting Centres in Nineteenth-Century Scientific Thinking* (London: Anthem Press, 2006); Gowan Dawson, *Darwin, Literature and Victorian Respectability* (Cambridge, U.K.: Cambridge University Press, 2007).

Chapter 4. Sympathy as Callousness? Physiology and Vivisection

1. *Spectator* 47 (1874), 13–14.

2. Peter J. Bowler, "Lankester, Sir (Edwin) Ray (1847–1929)," in *Oxford Dictionary of National Biography* (Oxford, U.K.: Oxford University Press, 2004), http://www.oxforddnb.com/view/article/34406, accessed February 1, 2012.

3. *Spectator* 47 (1874), 13–14.

4. Royal Commission on Vivisection, 128–29.

5. Ibid., 109.

6. Royal Commission of Vivisection, Report, x.

7. Royal Commission on Vivisection, 115–16, 144–46.

8. Ibid., 149, 156.

9. For a parallel case in Norway at the turn of the twentieth century, see Kristin Asdal, "Subjected to Parliament: The Laboratory of Experimental Medicine and the Animal Body," *Social Studies of Science* 38 (2008): 899–917.

10. The phrase "ghastly kitchen" ("affreuse cuisine") belongs to Claude Bernard's *Introduction à l'Étude de la Médecine Expérimentale* (Paris: J.B. Baillière et Fils, 1865), 28.

11. Patrizia Guarnieri, "Moritz Schiff (1823–96): Experimental Physiology and Noble Sentiment in Florence," in *Vivisection in Historical Perspective*, ed. Rupke, 106.

12. Stewart Richards, "Drawing the Life-Blood of Physiology: Vivisection and the Physiologists" Dilemma, 1870–1900," *Annals of Science* 43 (1996): 27–56 at 31, 47–48. See also

Hilda Kean, "'The Smooth Cool Men of Science': The Feminist and Socialist Response to Vivisection," *History Workshop Journal* 40 (1995): 16–38 at 19, 23.

13. Paul S. White, "The Experimental Animal in Victorian Britain," in Lorraine Daston and Gregg Mitman, eds., *Thinking with Animals: New Perspectives on Anthropomorphism* (New York: Columbia University Press, 2005), 62, 74. White's appraisal of the emotional control of physiologists is given richer treatment in "Sympathy under the Knife: Experimentation and Emotion in Late Victorian Medicine," in Fay Bound Alberti, ed., *Medicine, Emotion and Disease, 1700–1950* (Houndmills, U.K.: Palgrave, 2006), 100–24. See also Donald Fleming, "Charles Darwin, the Anaesthetic Man," *Victorian Studies* 4 (1961): 219–36.

14. See chapter 2. Inspiration for this analysis is taken from Monique Scheer, "Are emotions a kind of practice (and is that what makes them have a history)? A Bourdieuian approach to understanding," *History and Theory* 51 (2012): 193–220; and from Lorraine Daston, "Moral Economy." See also Thomas Schlich, "Surgery, Science and Modernity: Operating Rooms and Laboratories as Spaces of Control," *History of Science* 45 (2007): 231–56.

15. See also the dedication to Ludwig, "my revered and beloved master," in Stirling's *Outlines of Practical Physiology* (London: Charles Griffin, 1888).

16. C. Ludwig, "Die 'Vivisection' vor dem Richterstuhl der Gegenwart: Ein Wort zur Vermittelung," *Die Gartenlaube* (1879), 417–19; Monica Libell, *Morality beyond Humanity: Schopenhauer, Grysanowski, and Schweitzer on Animal Ethics* (Lund, Sweden: Ugglan, 2001), 241–43.

17. H. P. Bowditch, "The Physiological Laboratory at Leipzig," *Nature* 3 (1870): 142–43. Bowditch, who himself trained with Claude Bernard, would be the mentor of Walter B. Cannon at Harvard.

18. Quoted in Michael L. Geison, *Michael Foster and the Cambridge School of Physiology: The Scientific Enterprise in Late Victorian Society* (Princeton, N.J.: Princeton University Press, 1978), 162–69.

19. George D. Pollock, *An Address Delivered at St. George's Hospital on the Opening of the New Physiological Laboratory, 18th May 1887* (London: Adlard and Son, 1887), 15.

20. Dror, "The Scientific Image of Emotion"; "The Affect of Experiment."

21. Burdon-Sanderson, ed., *Handbook for the Physiological Laboratory*, II, plate LXXXIX, fig. 226; XCII, fig. 237; XCIII, fig. 242; CXII, fig. 308; CXIV, fig. 310; CXV, fig. 316.

22. Johann N. Czermak, *Über das physiologische Privat-Laboratorium an der Universität Leipzig: Rede gehalten am 21. December 1872, bei Gelegenheit der eröffnung seines Amphitheaters* (Leipzig, Germany: Wilhelm Engelmann, 1873), 10–18.

23. *Jackson's Oxford Journal*, March 14, 1885.

24. The term is William Reddy's. *The Navigation of Feeling*.

25. *Jackson's Oxford Journal*, March 14, 1885.

26. For a general description of the building, see *Jackson's Oxford Journal*, October 11, 1884. For the wider context of the debate, see Rob Boddice, "Vivisecting Major: A Victorian Gentleman Scientist Defends Animal Experimentation, 1876–1885," *Isis* 102 (2011): 215–37 at 221–26.

27. I omit a long list in favor of the principal secondary source: Lansbury, *The Old Brown Dog*.

28. Lind af Hageby and Leisa Katherina Schartau, *The Shambles of Science* (London: E. Bell, 1903), 19–26 (chapter entitled "Fun"), 20. "There is nothing of the serene dignity of science about the place, everybody looks as if he expected an hour's amusement" (21).

29. Royal Commission on Vivisection, 82–83. It was probably a version of Ferrier's Croonian Lecture. See David Ferrier, "The Croonian Lecture—Experiments on the Brain of Monkeys" (read by John Burdon-Sanderson), *Philosophical Transactions of the Royal Society of London*, 165 (1875): 433–88.

30. Steven Shapin, "The House of Experiment in Seventeenth-century England," *Isis* 79 (1988): 373–404, at 373, 404. See also Owen Hannaway, "Laboratory Design and the Aim of Science: Andreas Libavius versus Tycho Brahe," *Isis* 77 (1986): 584–610.

31. The classic study is George Weisz, *Divide and Conquer: A Comparative History of Medical Specialization* (Oxford, U.K.: Oxford University Press, 2006).

32. Boddice, "Vivisecting Major."

33. Bowditch, "The Physiological Laboratory," 142–43.

34. Royal Commission on Vivisection, 185.

35. Wellcome Library, London, Records of the Physiological Society, SA/PHY/C/1/1, March 31, 1876, 1–2.

36. Paul White, "Introduction," *Isis* 100 (2009): 792–97 at 796.

37. The best article on this subject is Stewart Richards, "Anaesthetics, ethics and aesthetics: Vivisection in the late nineteenth-century British laboratory," in Andrew Cunningham and Perry Williams, eds., *The laboratory revolution in medicine* (Cambridge, U.K.: Cambridge University Press, 1992), 142–69. For a positive account of how "saving pain" was "consonant with the very essence of modern civilization," see Stephanie J. Snow, *Blessed Days of Anaesthesia: How Anaesthetics Changed the World* (Oxford, U.K.: Oxford University Press, 2008), 155–58, 164.

38. Royal Commission on Vivisection, 32–34, 36.

39. See Royal Commission on Vivisection, testimony of Thomas Watson (4), William Sharpey (25–26), Alfred Swaine Taylor (60).

40. The *Times* (London), August 4, 1875.

41. Royal Commission on Vivisection, 12. The quotation is from the question put to Burrows by Viscount Cardwell.

42. Royal Commission on Vivisection, 17. The point was echoed by Alfred Swaine Taylor of Guy's Hospital (57). A fuller statement to the same effect was made by John Simon (75).

43. Royal Commission on Vivisection, 29. The point was reinforced by John Simon (75), and Phillip Henry Pye-Smith (109). The major dissention from this view came from Arthur de Noé Walker, who had witnessed vivisection predominantly without the presence of anesthetics, and mainly outside of Britain (esp. 246).

44. See, for example, George John Romanes, *Jelly-fish, Star-fish, and Sea-urchins: Being a Research on Primitive Nervous Systems* (London: Kegan Paul, Trench, 1885), 6–9.

45. E. Gurney, "A Chapter in the Ethics of Pain," *Fortnightly Review* 36 (1881): 778–96, 780, 783, 786–87. Charles Darwin was largely in agreement with Gurney. See Francis Darwin, ed., *The Life and Letters of Charles Darwin*, 3 vols. (London: John Murray, 1887), III, 210. See also W. Collier, "The Comparative Insensibility of Animals to Pain," *Nineteenth Century* 26 (October 1889): 622–27 at 622.

46. Collier, "Comparative Insensibility," 624.

47. John Stuart Mill, *Utilitarianism* (1863), in H. B. Acton, ed., *Utilitarianism, Liberty, Representative Government* (London and New York: Dent, 1972), 7–9; cf. Collier, "Comparative Insensibility," 623.

48. John Simon, "An Address delivered at the opening of the Section of Public Medicine," *BMJ*, August 6, 1881, 219–23 at 223. William Carpenter advised the physiologist to "never forget the pain he is inflicting, or lose sight of any means he can devise for its avoidance or mitigation," but with this in mind he wished the physiologist "'God Speed;' in the full conviction that his work is good and right." William B. Carpenter, "The Ethics of Vivisection," *Fortnightly Review* 31, no. 182 (1882): 237–46 at 46.

49. Emanuel Edward Klein was the most famous example of one who asserted that this was the only reason he used anesthetics, although he later recanted that assertion. See chapter 3, and Royal Commission on Vivisection, 183–85, 328. See also, Richards, "Drawing the Life-Blood of Physiology," 41; and Bruno Atalic and Stella Fatovic-Ferencic, "Emanuel Edward Klein—The Father of British Microbiology and the Case of the Animal Vivisection Controversy of 1875," *Toxicologic Pathology* 37 (2009): 708–13.

50. Carolyn Burdett, "Is Empathy the End of Sentimentality?" *Journal of Victorian Culture* 16 (2011): 259–74 at 269–70.

51. Osler, "Aequanimitas," 3–6.

52. The rise of physiology came late enough to adopt the post-mesmeric simplicity and objectification of chloroform: "One moment the patient was a conscious subject; the next, he or she was a body on the operating table." See Alison Winter, *Mesmerized: Powers of Mind in Victorian Britain* (Chicago: University of Chicago Press, 1998), 184.

53. Royal Commission on Vivisection, 282.

54. Cf. White, "Sympathy under the Knife," 112–14.

55. In addition to the examples given, see also George Fleming, "Vivisection and the Diseases of Animals," *Nineteenth Century* 11, no. 61 (1882): 468–78 (on the importance of vivisection to veterinary medicine); James Paget, Richard Owen, and Samuel Wilks, "Vivisection: Its Pains and Its Uses," *Nineteenth Century* 10, no. 58 (1881): 920–48.

56. Lady Burdon-Sanderson, *Sir John Burdon-Sanderson: A Memoir* (Oxford, U.K.: Clarendon Press, 1911), 157.

57. Royal Commission on Vivisection, 142.

58. Ibid., 146.

59. Ibid., 115, 119, 126.

60. Lady Burdon-Sanderson, *Sir John Burdon-Sanderson*, 101, 103.

61. Thomas Lauder Brunton, "Vivisection and the Use of Remedies," *Nineteenth Century* 11, no. 61 (1882): 479–87 at 479–80.

62. *BMJ*, August 13, 1881, 301.

63. Gerald F. Yeo, "The Practice of Vivisection in England," *Fortnightly Review* 31, no. 183 (1882): 352–68 at 358, 360; William W. Gull, "The Ethics of Vivisection," *Nineteenth Century* 11, no. 61 (1882): 456–67 at 458, 462, 466.

64. *Life and Letters of Charles Darwin*, III, 204. See also Francis Darwin, ed., *More Letters of Charles Darwin*, 2 vols. (London: John Murray, 1903), II, 435–41; Leonard Huxley, ed., *Life and Letters of Thomas Henry Huxley*, 2 vols. (London: MacMillan, 1900), I, 436–41.

65. On Romanes, Darwin, and Burdon-Sanderson, see Boddice, "Vivisecting Major," esp. 14; For Huxley, see Stephen Catlett, "Huxley, Hutton and the 'White Rage': A Debate

on Vivisection at the Metaphysical Society," *Archives of Natural History* 11 (1983): 181–89 at 185; Spencer was thus quoted in Albert Leffingwell, *An Ethical Problem: Or Sidelights upon Scientific Experimentation on Man and Animals* (London: G. Bell & Sons, 1916), 9.

66. Richards, *Darwin and the Emergence*, 121, 239.

67. The *Times* (London), April 18, 1881.

68. *Life and Letters of Charles Darwin*, III, 202.

69. Darwin, *Descent of Man*, 90.

70. Ibid., 159.

71. "Progress of Microscopical Science," *Monthly Microscopical Journal* 4 (1870): 226–27.

72. Huxley to Darwin, January 22, 1875 (dated January 18 according to Darwin Correspondence Project). Leonard Huxley, ed., *Life and Letters of Thomas Henry Huxley*, 3 vols. (London: Macmillan, 1913), II, 167.

73. Huxley to J. Donnelly, 1874. L. Huxley, ed., *Life and Letters* (1913), II, 159.

74. L. Huxley, *Life and Letters* (1913), II, 163.

75. Huxley to Darwin, October 30, 1875. L. Huxley, ed., *Life and Letters* (1913), II, 172.

76. For details see L. Huxley, *Life and Letters* (1913), II, 157–58.

77. Huxley to Donnelly, 1874. L. Huxley, *Life and Letters* (1913), II, 159.

78. Ibid.

79. T. H. Huxley, *Lessons in Elementary Physiology* (London: Macmillan, 1866).

80. Ibid., 2.

81. Huxley, *Lessons*, 2nd ed. (1868).

82. Huxley, *Lessons*, 6th ed. (1872), v.

83. L. Huxley, ed., *Life and Letters* (1913), II, 153.

84. The *Times* (London), May 26, 1876; L. Huxley, *Life and Letters* (1913), II, 156–7.

85. Huxley to Donnelly, 1874. L. Huxley, *Life and Letters* (1913), II, 160.

86. T. H. Huxley, "The Progress of Science" (1887), *Collected Essays*, I, 122 *seq*. L. Huxley, *Life and Letters* (1913), II, 163–64.

87. See Rob Boddice, "The Moral Status of Animals and the Historical Human Cachet," *JAC* 30 (2010): 457–89 at 469–75.

88. T. H. Huxley, "The Progress of Science" (1887), *Collected Essays*, I, 122 *seq*. L. Huxley, *Life and Letters* (1913), II, 163–64.

89. Huxley to "A Student," 1890. L. Huxley, *Life and Letters* (1913), II, 165–66.

90. Boddice, "Vivisecting Major."

91. George John Romanes, *Christian Prayer and General Laws, being the Burney Prize Essay for the Year 1873, with an Appendix, the Physical Efficacy of Prayer* (London: Macmillan, 1874), 124.

92. Physicus (George John Romanes), *A Candid Examination of Theism* (London: Trübner, 1878), 113–14.

Chapter 5. Sympathy, Liberty, and Compulsion: Vaccination

1. Thomas Chalmers, *The Adaptation of External Nature to the Moral and Intellectual Constitution of Man, Treatise 1 of the Bridgewater Treatises on the Power, Wisdom, and Goodness of God, as Manifested in the Creation*, 2 vols. (London: William Pickering, 1833, 1839), II, 18–19.

2. Martha Nussbaum, *Upheavals of Thought: The Intelligence of Emotions* (Cambridge, U.K.: Cambridge University Press, 2003), 403–4. See also Laurent Berlant, "Introduction: Compassion (and Withholding)," in *Compassion: The Culture and Politics of Emotion*, ed. Laurent Berlant (New York: Routledge, 2004), 1–4.

3. For an overview of the *longue durée* controversy surrounding the introduction of vaccination, see Gareth Williams, *Angel of Death: The Story of Smallpox* (Houndmills, U.K.: Palgrave MacMillan, 2010), esp. chapters 8–13. For Jenner, see Rob Boddice, *Edward Jenner* (Stroud, U.K.: The History Press, 2015). For more specifically on the Victorian case, see Alison Bashford, *Imperial Hygiene: A Critical History of Colonialism, Nationalism and Public Health* (Houndmills, U.K.: Palgrave, 2003).

4. There is an impressive body of work on the development of smallpox vaccination in Britain, ranging from the social historical to the medical historical, to the cultural historical. It has been viewed as an exemplar of class tyranny (particularly by Nadja Durbach) and as a forerunner of the National Health Service. See in particular Nadja Durbach, *Bodily Matters: The Anti-Vaccination Movement in England, 1853–1907* (Durham, N.C.: Duke University Press, 2005); Stanley Williamson, *The Vaccination Controversy: The Rise, Reign and Fall of Compulsory Vaccination for Smallpox* (Liverpool, U.K.: Liverpool University Press, 2007); Deborah Brunton, *The Politics of Vaccination: Practice and Policy in England, Wales, Ireland, and Scotland, 1800–1874* (Rochester, N.Y.: University of Rochester Press, 2008); Dorothy Porter and Roy Porter, "The Politics of Anti-vaccinationism and Public Health in Nineteenth-Century England," *Medical History* 32 (1988): 231–52; Nadja Durbach, "'They Might As Well Brand Us': Working-Class Resistance to Compulsory Vaccination in Victorian England," *Social History of Medicine* 13 (2000): 45–62.

5. Vaccination Act, 16 & 17 Vict. c. 100, 1853; 30 & 31 Vict. c. 84, 1867; 34 & 35 Vict. c. 98, 1871. See also U.K. Parliament, 1871 (246) Report from the Select Committee on the Vaccination Act (1867), together with the proceedings of the committee, minutes of evidence, appendix, and index. For the acknowledged human impact of this new organization, see Simon Szreter, "The Importance of Social Intervention in Britain's Mortality Decline *c.* 1850–1914: A Re-interpretation of the Role of Public Health," *Social History of Medicine* 1 (1988): 1–38. For a more specific argument, see R. J. Lambert, "A Victorian National Health Service: State Vaccination 1855–71," *Historical Journal* 5 (1962): 1–18. For the centralization of sanitary and public health administration brought about by the vaccination laws, see Graham Mooney, "'A Tissue of the Most Flagrant Anomalies': Smallpox Vaccination and the Centralization of Sanitary Administration in Nineteenth-Century London," *Medical History* 41 (1997): 261–90.

6. See Sally Sheard and Liam J. Donaldson, *The Nation's Doctor: The Role of the Chief Medical Officer 1855–1998* (Abingdon, U.K.: Radcliffe, 2006), 1–15 for a brief overview. The principal biography is R. Lambert, *Sir John Simon, 1816–1904 and English Social Administration* (London: MacGibbon & Kee, 1963).

7. Lambert, *Sir John Simon*, 256, 272, 324–25, 328, 392, 446.

8. *The Times* (London), September 20, 1875.

9. Lambert, *Sir John Simon*, 302, 486.

10. Judith Rowbotham, "Legislating for Your Own Good: Criminalising Moral Choice. The Modern Echoes of the Victorian Vaccination Acts," *Liverpool Law Review* 30 (2009): 13–33, at 16.

11. For the Leicester case, see Stuart M. F. Fraser, "Leicester and Smallpox: The Leicester Method," *Medical History* 24 (1980): 315–32; Dale-L. Ross, "Leicester and the Anti-vaccination Movement, 1853–1889," *Leicestershire Archaeological and Historical Society* 43 (1967–1968): 35–44; J. D. Swales, "The Leicester Anti-Vaccination movement," *The Lancet* 340, no. 8826 (1992): 1019–21.

12. There is, of course, an ongoing contemporary connection between fear and vaccination of all kinds. See Andrea Kitta, *Vaccinations and Public Concern in History: Legend, Rumor, and Risk Perception* (New York and London: Routledge, 2011); Rob Boddice, "Vaccination, Fear and Historical Relevance," *History Compass* 14 (2016): 71–78.

13. U.K. Parliament, Royal Commission Appointed to Inquire into the Subject of Vaccination (hereafter Vaccination Commission), First Report, C. 5845 (1889); Second Report, C. 6066 (1890); Third Report, C. 6192 (1890); Fourth Report, C. 6527 (1890–1891); Fifth Report, C. 6666 (1892); Sixth Report, C. 7993 (1896); Final Report, C. 8270 (1896). See also 61 & 62 Vict. c. 49, 1898, for the inclusion of a conscientious objection clause. The conscience clause was more effectively applied in a further amendment, 7 Edw. VII, c. 31, 1907.

14. A secondary question was whether vaccination was medically effective. The debate about the medical efficacy of vaccination will detain us here, only where it impinges on the principal question of the ethics of administering it under compulsion.

15. Darwin, *Descent of Man*, 159. Seemingly only one anti-vaccinationist correspondent of Darwin's took exception to his view, which Darwin evidently defended (although his side of the correspondence has apparently not survived). See J. M. Grandclément to Darwin, May and June 1874, Darwin Correspondence Database, http://www.darwinproject.ac.uk/entry- 9436 and - 9479. Darwin also entertained the idea, being still committed to a belief in the inheritance of acquired characteristics, of the inheritance of acquired immunity, put forward by Alphonse de Candolle in *Histoire des sciences et des savants depuis deux siècles* (Geneva: H. Georg, 1873), 427–31. Candolle had put this forward as a new application of Darwin's ideas, suggesting that a lack of exposure to smallpox itself in the generation after Jenner's discovery had led to less latent immunity in the population. Where vaccination worked well in the population with a certain amount of inherited immunity, it would work less well after several generations, because selection for smallpox immunity would have ceased. A population re-exposed to the disease at this point would fare relatively poorly. Darwin in turn told Candolle that his chapter on vaccination was the most interesting thing in the book. Darwin to Candolle, December 11, 1872, Darwin Correspondence Project, http://www.darwinproject.ac.uk/entry-8672.

16. John Simon, *English Sanitary Institutions*, 2nd ed. (London: John Murray, 1897), 311–13.

17. Darwin, *Descent of Man*, 64–65.

18. Charles Darwin, *Life of Erasmus Darwin* (1879), ed. Desmond King-Hele (Cambridge, U.K.: Cambridge University Press, 2003), 83.

19. Cambridge University Library, DAR112.B3f.

20. Mill to Spencer, July 30, 1861. Quoted in Richards, *Darwin and the Emergence*, 300.

21. On sympathy and the evolution of the moral sense, see Herbert Spencer, *The Principles of Ethics*, 2 vols. (1892; New York: D. Appleton, 1897–1898), I, vi; for his anti-vaccinationism on the grounds of liberty, see *Vaccination Inquirer* 2, no. 13 (1880): 9; on

his doubts about vaccination's efficacy, see Herbert Spencer, "Vaccination," *Facts and Comments* (New York: D. Appleton, 1902), 270–73; Herbert Spencer, *Social Statics or Order* (1892; New York: D. Appleton, 1915), 212–18, esp. 212–13; and Herbert Spencer, "The Man Versus the State" (1884), in *The Man Versus the State, with six Essays on Government, Society, and Freedom* (Indianapolis: Liberty Classics, 1981), 17–20.

22. Dixon, *Invention of Altruism*, 219.

23. Herbert Spencer, *The Study of Sociology* (New York: D. Appleton, 1874), 360–61.

24. *Social Statics*, 212–18; Spencer, "Man Versus the State," 30.

25. Spencer, "Vaccination," 270.

26. Spencer, *Principles of Ethics*, II, 392, 357. Thomas Dixon justly notes the sternness of Spencer's "prioritization of the health of society over the health of the individual." See Dixon, *Invention of Altruism*, 208.

27. Huxley to Spencer, January 18, 1887. L. Huxley, *Life and Letters* (1913), III, 49–50.

28. T. H. Huxley, "Administrative Nihilism" (1871), *Method and Results: Essays* (New York: D. Appleton, 1898), 261–63.

29. Joseph Lister, "On the Interdependence of Science and the Healing Art" (presidential address to the British Association for the Advancement of Science, Liverpool, 1896), in *The Collected Papers of Joseph, Baron Lister* (Oxford, U.K.: Clarendon, 1909), II, 505–6.

30. A. F. Vulliamy, *Small Pox and Vaccination* (Ipswich, U.K.: Norman Adlard, 1896), 12.

31. T. H. Huxley, "Yeast," lecture delivered in Manchester, November 13, 1871, printed in *Contemporary Review*, 1871. The argument was re-employed to defend vivisection, which technically included Pasteur's attempts to generate vaccines for other diseases. See "Vivisection and Medicine," *Nature* 24 (1881): 329–32.

32. T. H. Huxley, speech on "Spontaneous Generation," September 14, 1870, printed in *Nature* 2 (1870): 400–406. Passage on vaccination at 404. Quotation at 406. Reprinted in *Critiques and Addresses* as "Biogenesis and Abiogenesis" (London: Macmillan, 1890).

33. The sentiment was shared by Edwin Lankester (father of E. Ray Lankester), who was a great friend of Huxley's and Darwin's, and prominent researcher and promoter of public health. See E. Lankester, "The Small-Pox Epidemic," *Nature* 3 (1871): 341–42. Lankester recommended house-to-house visitations of public vaccinators across London, with the power to compel vaccination or re-vaccination.

34. *London Daily News*, August 3, 1880; for the measure, see 19th Century House of Commons Sessional Papers, vol. VII, 595 (1880).

35. *London Standard*, July 27, 1880; *BMJ* (1880), vol. 2, 178–81.

36. *BMJ* (1880), vol. 2, 179–80.

37. In 1871 Huxley was accused of being an anti-vaccinationist in the *Marylebone Mercury*, which caused *The Lancet* to seek clarification from the man himself. Huxley's reply to *The Lancet* was relayed thus: "during his candidature for a seat on the School Board he found himself compelled to contradict a report, which had been industriously set about, that he was a champion of spirit-rapping and table-turning [spiritualism and séances]; and he cannot but think that the report that he is opposed to compulsory vaccination is a piece of work from the same manufactory." *The Lancet* felt inclined to "rejoice in being able to meet it [the accusation] with an authoritative and formal contradiction." *The Lancet* 97, no. 2476 (1871): 204.

38. For an overview of Wallace's activities at this time, see Ross A. Slotten, *The Heretic in Darwin's Court: The Life of Alfred Russel Wallace* (New York: Columbia University Press, 2006), 422–55. For Wallace's anti-vaccinationism, see Martin Fichman, "Alfred Russel Wallace and Anti-vaccinationism in the Late Victorian Cultural Context, 1870–1907," in *Natural Selection and Beyond: The Intellectual Legacy of Alfred Russel Wallace,* ed. Charles H. Smith and George Beccaloni (Oxford, U.K.: Oxford University Press, 2008); and Martin Fichman and Jennifer E. Keelan, "Resister's Logic: The Anti-Vaccination Arguments of Alfred Russel Wallace and Their Role in the Debates Over Compulsory Vaccination in England, 1870–1907," *Studies in History and Philosophy of Biological and Biomedical Sciences* 38 (2007): 585–607. For Wallace's departures from the mainstream of scientific opinion, see, for example, Fern Elsdon-Baker's account of the reception of Wallace's spiritualism: "Spirited Dispute: The Secret Split between Wallace and Romanes," *Endeavour* 32 (2008): 75–78.

39. Alfred Russel Wallace, "The Limits of Natural Selection as Applied to Man," *Contributions to the Theory of Natural Selection: A Series of Essays* (London and New York: MacMillan, 1870); Alfred Russel Wallace, "How to Nationalize the Land: A Radical Solution of the Irish Land Problem," *Contemporary Review* 38 (1880): 716–36; Alfred Russel Wallace, *To Members of Parliament and Others: Forty-five Years of Registration Statistics, Proving Vaccination To Be Both Useless and Dangerous* (London: E.W. Allen, 1885).

40. Alfred Russel Wallace, "The Development of Human Races under the Law of Natural Selection," in *Contributions to the Theory of Natural Selection,* 328.

41. *Vaccination Inquirer* (1911), quoted in J. T. Biggs, *Leicester: Sanitation versus Vaccination* (London: National Anti-vaccination League, 1912), 624.

42. Vaccination Commission, Third Report, 23.

43. Ibid., 20.

44. Ibid., 21.

45. Ibid., 13, 127.

46. Wallace was extolling the virtues of a new socialist magazine, *The Eagle and the Serpent,* in *Clarion* 328 (March 19, 1898): 95.

47. Vaccination Commission, Third Report, 25.

48. Biggs, *Leicester,* 624.

49. Vaccination Commission, Third Report, 127.

50. For Wallace's difficulty in squaring away this kind of compulsion, see Vaccination Commission, Third Report, 128–29, esp. questions 9667–72; for the discovery of the vaccinated cordon in Leicester, see *The Lancet* 127, no. 3275, 1090–91.

51. Wallace to Prof. Barrett, October 30, 1899, in James Marchant, *Alfred Russel Wallace: Letters and Reminiscences* (London: Cassell, 1916), vol. 2, 206.

52. Myers to Wallace, April 12, 1890, British Library, Alfred Russel Wallace papers, add. MS 46440, f.330.

53. Alfred Russel Wallace, *Vaccination A Delusion, Its Enforcement A Crime: Proved by the official evidence in the reports of the Royal Commission* (London: Swan Sonnenschein, 1898), 92. Most of the letters in British Library, add. MSS 46440, ff. 293–353 are concerned with the preparation and distribution of this text.

54. See T. H. Huxley, "Evolution and Ethics: Prolegomena," *Collected Essays* 9, *Evolution & Ethics and Other Essays* (London: MacMillan, 1894), 30–39 and chapter 6, below.

55. *The Lancet* 151, no. 3891 (1898): 894.

Chapter 6. Sympathetic Selection: Eugenics

1. Diane B. Paul and James Moore, "The Darwinian Context: Evolution and Inheritance," in *The Oxford Handbook of the History of Eugenics*, ed. Alison Bashford and Philippa Levine (Oxford, U.K.: Oxford University Press, 2010), 27–42, at 28. For the general context of Darwinism's influence on early eugenicists, see Daniel Pick, *Faces of Degeneration: A European Disorder, c.1848-c.1918* (Cambridge, U.K.: Cambridge University Press, 1989), 216–21.

2. See, for example, the compendious volume edited by Bashford and Levine, *The Oxford Handbook of the History of Eugenics*. For further reviews of the field, see Mathew Thomson, *The Problem of Mental Deficiency: Eugenics, Democracy, and Social Policy in Britain c.1870–1959* (Oxford, U.K.: Clarendon Press, 1998); Peter Bowler, *The Mendelian Revolution: The Emergence of Hereditarian Concepts in Modern Science and Society* (London: Athlone, 1989); G. R. Searle, *Eugenics and Politics in Britain, 1900–1914* (Leyden, Netherlands: Noordhoff International, 1976); Pauline M.H. Mazumdar, *Eugenics, Human Genetics and Human Failings: The Eugenics Society, its Sources and its Critics in Britain* (London: Routledge, 1992); Richard Soloway, *Demography and Degeneration: Eugenics and the Declining Birthrate in Twentieth-century Britain* (Chapel Hill: University of North Carolina Press, 1990); Marius Turda, "Recent Scholarship on Race and Eugenics," *Historical Journal* 51 (2008): 1115–24; Frank Dikötter, "Race Culture: Recent Perspectives on the History of Eugenics," *American Historical Review* 103 (1998): 467–78; Robert A. Nye, "The Rise and Fall of the Eugenics Empire: Recent Perspectives on the Impact of Biomedical Thought in Modern Society," *Historical Journal* 36 (1993): 687–700.

3. Marius Turda, *Modernism and Eugenics* (Houndmills, U.K.: Palgrave Macmillan, 2010), 3.

4. This is slightly against the grain, because the emphasis of late has been on comparative approaches to the history of eugenics, where well-formulated narratives of national histories are juxtaposed (see Philippa Levine and Alison Bashford, "Introduction: Eugenics and the Modern World," in *The Oxford Handbook of the History of Eugenics*, 3–24, at 20). I am picking a little bit at the seams of one of those well-formulated narratives in an attempt to show that a new approach can reveal that we did not understand the origins of eugenics in Britain quite as well as we thought we did.

5. The hope was explicit: "but there appears at least one check in steady action, namely that the weaker and inferior members of society do not marry so freely as the sound; and this check might be indefinitely increased by the weak in body or mind refraining from marriage, though this is more to be hoped for than expected" (*Descent of Man*, 159–60).

6. James Moore notes the irresistible suggestion of eugenics in Darwin's *Descent of Man*. See "R. A. Fisher: A faith fit for eugenics," *Studies in History and Philosophy of Biological and Biomedical Sciences*, 38 (2007): 110–35, at 113.

7. See chapter 5, note 24.

8. Spencer, *Principles of Ethics*, II, 358.

9. Spencer, *Principles of Psychology*, II, 688–692. Spencer draws explicitly on Alexander Bain in this passage. See Bain, *Emotions and the Will*, 3rd ed. (London: Longmans, Green, 1875), 142–44, 179.

10. David Konstan, *Pity Transformed* (London: Duckworth, 2001), 12.

11. Himmelfarb, *Poverty and Compassion*, 197–98.

12. See Dixon, *Invention of Altruism*, 196–97.

13. Herbert Spencer, "Morals and Moral Sentiments," *Fortnightly Review* 52 (1871): 419–32 at 432.

14. Cf. Peter J. Bowler, *Biology and Social Thought: 1850–1914* (Berkeley: Office for History of Science and Technology, University of California, 1993), 79–80. Bowler pegs Galton as a saltationist, for whom eugenics "could not be a form of social Darwinism since it did not draw any analogy between the manipulation of human heredity and the process of natural evolution." The "next step in human evolution," Bowler states, "would have to wait for the appearance of a suitable saltation." The argument sustained throughout this book is that evolutionary scientists, who had had the extraordinary capacity of understanding evolutionary processes, were themselves such a "saltation."

15. Francis Galton, *Hereditary Genius: An Inquiry into Its Laws and Consequences* (London: Macmillan, 1869), 356–57.

16. Francis Galton, "Hereditary Improvement," *Fraser's Magazine* 7 (1873): 116–30 at 119.

17. Francis Galton, *Inquiries into Human Faculty*, 27.

18. Galton, "Hereditary Improvement," 120.

19. Francis Galton, "Eugenics: Its Definition, Scope and Aims," *Sociological Papers* 1 (1905): 45–50 at 50.

20. Galton, *Hereditary Genius*, 358–59.

21. Francis Galton, *English Men of Science: Their Nature and Nurture* (London: Macmillan, 1874), 260.

22. Karl Pearson, *The Life, Letters and Labours of Francis Galton*, 4 vols. (Cambridge, U.K.: Cambridge University Press, 1914–30), II, 249.

23. Pearson, *Life*, II, 261. See also Daniel J. Kevles, *In the Name of Eugenics: Genetics and the Uses of Human Heredity* (Berkeley and Los Angeles: University of California Press, 1985), 12.

24. Galton, *Inquiries into Human Faculty*, 300.

25. Ibid., 304

26. Ibid., 334–35.

27. Galton's Herbert Spencer Lecture of 1907, quoted in D. W. Forrest, *Francis Galton: The Life and Work of a Victorian Genius* (London: Paul Elek, 1974), 272–73. The cultural emphasis is in keeping with the general thrust of late Victorian race science. See Douglas A. Lorimer, *Science, Race Relations and Resistance: Britain, 1870–1914* (Manchester, U.K.: Manchester University Press, 2013), 59–99.

28. Francis Galton, "Eugenics as a Factor in Religion," *Sociological Papers* 2 (1906): 52–53 at 53. The idea of a new "nobility" in the context of eugenics was also advocated by R. A. Fisher in 1912. See Moore, "R.A. Fisher," 121.

29. Galton, *Inquiries into Human Faculty*, 24–25*n*.

30. Ibid., 336.

31. Francis Galton, "The Possible Improvement of the Human Breed under Existing Conditions of Law and Sentiment," *Essays in Eugenics* (1909; New York: Garland, 1985), 30 (delivered as the second Huxley lecture, October 29, 1901. This note was added to the text for the 1909 printing).

32. Galton in the *Westminster Gazette*, June 26, 1908, quoted in Forrest, *Francis Galton*, 276.

33. Galton, "Hereditary Improvement," 123.

34. See Galton, "Hereditary Improvement," 125–26; and Pearson, *Life*, IIIa, 292–96 (Galton's unpublished manuscript On Eugenic Certificates [1906]).

35. C. W. Saleeby, "The Psychology of Parenthood," *Eugenics Review* 1 (1909–1910): 37–46 at 41–43.

36. See Stephanie Olsen, *Juvenile Nation: Youth, Emotions and the Making of the Modern British Citizen, 1880–1914* (London: Bloomsbury, 2014), esp. chapter 6.

37. Pearson, *Life*, IIIa, 411.

38. Ibid., 413.

39. Francis Galton, *Memories of My Life* (London: Methuen, 1908), 321–23.

40. Francis Galton, "Eugenic Qualities of Primary Importance," *Eugenics Review* 1 (1909–1910): 74–76 at 76.

41. Pearson's modern biographer (Theodore M. Porter, *Karl Pearson: The Scientific Life in a Statistical Age* [Princeton, N.J.: Princeton University Press, 2005]) rather passes over the centrality of Pearson's racism to both his politics and his scientific practice. See, however, 278f for an overview of Pearson's correlation of eugenics, evolution, and statistics, including the observation, "Eugenics would not halt the advance of the finer moral sympathies, but guide them along the paths of efficiency and racial progress" (282).

42. See Kevles, *In the Name of Eugenics*, 24: "Pearson was concerned less with the shape of the new society than with where the Karl Pearsons would fit into it."

43. Karl Pearson, *Social Problems: Their Treatment, Past, Present, and Future* (London: Dulau, 1912), 11, 15.

44. Karl Pearson, *The Academic Aspect of the Science of National Eugenics: A Lecture Delivered to Undergraduates* (London: Dulau, 1911), 18–19.

45. Karl Pearson, *The Groundwork of Eugenics* (London: Dulau. 1909), 23.

46. Karl Pearson, *National Life from the Standpoint of Science* (London: Adam and Charles Black, 1901), 46.

47. Ibid., 58–59.

48. Ibid., 61–62.

49. Karl Pearson, *The Grammar of Science* (London: Walter Scott, 1892), 432–39.

50. Pearson, *National Life*, 54.

51. Pearson, *Academic Aspect*, 18–19; Karl Pearson, *Darwinism, Medical Progress, and Eugenics: The Cavendish Lecture, 1912* (London: Dulau, 1912), 21.

52. For Huxley's prediction see T. H. Huxley, "Evolution and Ethics: Prolegomena," 30–39, and see below. For "medical mathematicians" see Pearson, *Darwinism*, 9–11. See also Kevles for the rationale of Pearson's objection, for example, to the Factory Acts. "The prohibitions against child labor transformed children into economic liabilities, and the better class of workers quickly reduced their birth rate, leaving the principal task of procreation to the socially worst" (*In the Name of Eugenics*, 33). For Pearson's own expression of this rational sympathy see Karl Pearson, *The Problem of Practical Eugenics* (London: Dulau, 1912), 22–25.

53. Karl Pearson, *On the Scope and Importance to the State of the Science of National Eugenics* (London: Dulau, 1909), 24–25.

54. Pearson, *On the Scope*, 37–38.
55. Karl Pearson, ed., *Treasury of Human Inheritance* (London: Dulau, 1912), I, iii.
56. Pearson, *Grammar of Science*, 42–43.
57. For Weldon's relationship with Pearson, see Kevles, *In the Name of Eugenics*, 27–30, 35–36.
58. Pearson to Galton, January 24, 1906, Pearson, *Life*, IIIa, 278–79.
59. Galton to Pearson, February 1, 1906, Pearson, *Life*, IIIa, 280.
60. Galton to Pearson, April 16, 1906, Pearson, *Life*, IIIa, 280.
61. See "The scope of Biometrika," *Biometrika* 1 (1901): 1–2.
62. Pearson, *Life*, IIIa, 281.
63. Kevles, *In the Name of Eugenics*, 36.
64. [Karl Pearson], "Walter Frank Raphael Weldon. 1860–1906," *Biometrika* 5 (1906): 1–52 at 1–2.
65. Pearson, *Life*, IIIa, 441–42.
66. For Galton's clearest expression of this, see "The Part of Religion in Human Evolution," *National Review*, August 1894, 755–65. See also Galton's "Restrictions in Marriage," *Sociological Papers* 2 (1906): 3–13, and the appended essay "Eugenics as a Factor in Religion," 52–53.
67. Pearson, *Life*, IIIa, 88–89. The association was too much for many. See Moore, "R.A. Fisher," 117. Leonard Darwin (Charles's youngest son) maintained the importance of religion, "including all the promptings of the inner man toward better things," during his presidency of the Eugenics Education Society, when he replaced Galton in that position. See L. Darwin, *What is Eugenics?* Special ed. (New York: Third International Congress of Eugenics, 1932), 88.
68. Pearson, *Life*, IIIa, 442.
69. E. S. Pearson, "Karl Pearson: An Appreciation of Some Aspects of His Life and Work," *Biometrika* 28 (1936): 193–256 at 194.
70. Kevles, *In the Name of Eugenics*, 25–26. Eugenic marriage was far from uncommon, especially among "new women." See Angelique Richardson, *Love and Eugenics in the Late Nineteenth Century: Rational Reproduction and the New Woman* (Oxford, U.K.: Oxford University Press, 2003).
71. T. H. Huxley, "Evolution and Ethics: Prolegomena," 30–39.
72. Vaccination Commission, Third Report, 25.
73. John Simon, "On the Ethical Relations of Early Man," *Nineteenth Century* 206 (1894), reprinted in *English Sanitary Institutions*, 2nd ed. (London: John Murray, 1897), 489–97.
74. Ibid., 491.
75. Simon, *English Sanitary Institutions*, xii.
76. Simon, "Ethical Relations," 497.
77. Simon, *English Sanitary Institutions*, 449.
78. Ibid., 485.
79. Ibid., 488.
80. J. Arthur Thomson, *Darwinism and Human Life* (London: Andrew Melrose, 1909), 213–14.
81. Ibid., 217–18.
82. Ibid., 219–20.

83. Ibid., 230, 234.

84. The war was in part responsible for the government investigation that led to the *Report of the Inter-Departmental Committee on Physical Deterioration* (1904). The third edition of Booth's works were published in seventeen volumes, London: MacMillan, 1902–1903.

85. See Thomas R. C. Brydon, "Charles Booth, Charity Control, and the London Churches, 1897–1903," *The Historian* 68 (2006): 489–518; Matthew Hilton, *Smoking in British Popular Culture, 1800–2000: Perfect Pleasures* (Manchester, U.K.: Manchester University Press, 2000), 60–80; Olsen, *Juvenile Nation*, esp. ch. 6; and G. R. Searle, *The Quest for National Efficiency: A Study in British Politics and Political Thought, 1899–1914* (Berkeley and Los Angeles: University of California Press, 1971).

Conclusion

1. Romanes to Huxley, January 3, 1879, Huxley Papers, 25.206-7, Imperial College, London.

Bibliography

This list is not a comprehensive compilation of every source used in the book, but rather a guide to both secondary reading in the history of emotions as it intersects with the histories of science and morality, especially in the Victorian period, and to easily accessible publications from the historical period in question. As such, this list serves as a framework both for following up the arguments made in *Science of Sympathy* and as a route-map for further study.

Published Primary Sources

Bain, Alexander. *Emotions and the Will.* London: John W. Parker and Son, 1859.

———. *The Senses and the Intellect,* 4th ed. London: Longmans, Green, 1894.

Bell, Charles. *Essays on the Anatomy of Expression in Painting.* London: Longman, Hurst, Rees, and Orme, 1806.

Burdon-Sanderson, John, ed. *Handbook for the Physiological Laboratory,* by E. Klein, John Burdon-Sanderson, Michael Foster, and Thomas Lauder Brunton, 2 vols. London: J & A Churchill, 1873.

Cobbe, Frances Power. *Life of Frances Power Cobbe.* Boston and New York: Houghton, Mifflin and Company, 1904.

———. *Re-Echoes.* London: Williams and Norgate, 1876.

———. *Darwinism in Morals, and Other Essays.* London: Williams and Norgate, 1872.

Darwin, Charles. *Autobiographies.* London: Penguin, 2002.

———. *The Descent of Man, and Selection in Relation to Sex.* 1871; London: Penguin, 2004.

———. *The Expression of Emotions in Man and Animals.* London: John Murray, 1872.

———. *The Origin of Species by Means of Natural Selection or The Preservation of Favoured Races in the Struggle for Life.* 1859; London: Penguin, 1985.

Galton, Francis. *English Men of Science: Their Nature and Nurture.* London: MacMillan, 1874.

———. *Hereditary Genius: An Inquiry into Its Laws and Consequences.* London: MacMillan, 1869.
———. *Inquiries into Human Faculty and its Development.* London: MacMillan, 1883.
———. *Memories of My Life.* London: Methuen, 1908.
Huxley, Leonard, ed. *Life and Letters of Thomas Henry Huxley,* 2nd ed., 3 vols. London: MacMillan, 1913.
Huxley, T. H. *Collected Essays.* London: MacMillan, 1894.
———. *Method and Results: Essays.* New York: D. Appleton, 1898.
Kidd, Benjamin. *Social Evolution.* London: MacMillan, 1894.
Osler, William. *Aequanimitas, with other Addresses to Medical Students and Practitioners of Medicine,* 2nd ed. Philadelphia: P. Blakiston's Son, 1925.
Paget, James, Richard Owen, and Samuel Wilks. "Vivisection: Its Pains and Its Uses," *Nineteenth Century* 10 (1881): 902–48.
Pearson, Karl. *Darwinism, Medical Progress, and Eugenics: The Cavendish Lecture, 1912.* London: Dulau, 1912.
———. *National Life from the Standpoint of Science.* London: Adam and Charles Black, 1901.
———. *On the Scope and Importance to the State of the Science of National Eugenics.* London: Dulau, 1909.
———. *Social Problems: Their Treatment, Past, Present, and Future.* London: Dulau, 1912.
———. *The Academic Aspect of the Science of National Eugenics: A Lecture Delivered to Undergraduates.* London: Dulau, 1911.
———. *The Grammar of Science.* London: Walter Scott, 1892.
———. *The Groundwork of Eugenics.* London: Dulau, 1909.
———. *The Life, Letters and Labours of Francis Galton,* 4 vols. Cambridge, U.K.: Cambridge University Press, 1914–30.
———. *The Problem of Practical Eugenics.* London: Dulau, 1912.
Pearson, Karl, ed., *Treasury of Human Inheritance.* London: Dulau, 1912.
Romanes, George John. *Animal Intelligence,* 3rd ed. London: Kegan Paul, Trench, 1882.
———. *Mental Evolution in Animals.* London: Kegan Paul, Trench, 1883.
———. *Mental Evolution in Man: Origin of Human Faculty.* London: Kegan Paul, Trench, 1888.
Simon, John. *English Sanitary Institutions,* 2nd ed. London: John Murray, 1897.
Smith, Adam. *The Theory of Moral Sentiments,* 1759. London: Penguin, 2009.
Spencer, Herbert. *Principles of Biology.* London: Williams & Norgate, 1864.
———. *Social Statics or Order.* 1892. New York: D. Appleton, 1915.
———. *Principles of Psychology.* 2 vols. 1899. Osnabrück, Germany: Otto Zeller, 1966.
———. *The Man Versus the State, with six Essays on Government, Society, and Freedom.* Indianapolis: Liberty Classics, 1981.
———. *The Principles of Ethics,* 2 vols. 1892. New York: D. Appleton, 1897–1898.
———. *The Study of Sociology.* New York: D. Appleton, 1874.
Thomson, J. Arthur. *Darwinism and Human Life.* London: Andrew Melrose, 1909.
U.K. Parliament. Report of the Royal Commission on the Practice of Subjecting Live Animals to Experiments for Scientific Purposes, C. 1397 (1876).
———. Royal Commission Appointed to Inquire into the Subject of Vaccination, First Report, C. 5845 (1889); Second Report, C. 6066 (1890); Third Report, C. 6192 (1890);

Fourth Report, C. 6527 (1890–1); Fifth Report, C. 6666 (1892); Sixth Report, C. 7993 (1896); Final Report, C. 8270 (1896).

Wallace, Alfred Russel. "The Development of Human Races under the Law of Natural Selection." In *Contributions to the Theory of Natural Selection: A Series of Essays*. London and New York: MacMillan, 1870.

———. *To Members of Parliament and Others: Forty-five Years of Registration Statistics, Proving Vaccination To Be Both Useless and Dangerous*. London: E.W. Allen, 1885.

———. *My Life*, 2 vols. New York: Dodd, Mead, 1905.

Wells, H. G. *The Island of Dr. Moreau*, 1896. London: Heinemann, 1921.

Wilde, Oscar. *The Picture of Dorian Gray*, 1891. New York: Mondial, 2015.

Secondary Sources

Alberti, Fay Bound. *Matters of the Heart: History, Medicine and Emotion*. Oxford, U.K.: Oxford University Press, 2010.

Asdal, Kristin. "Subjected to Parliament: The Laboratory of Experimental Medicine and the Animal Body." *Social Studies of Science* 38 (2008): 899–917.

Bashford, Alison. *Imperial Hygiene: A Critical History of Colonialism, Nationalism and Public Health*. Houndmills, U.K.: Palgrave, 2003.

Bittel, Carla. "Science, Suffrage, and Experimentation: Mary Putnam Jacobi and the Controversy over Vivisection in Late Nineteenth-Century America." *Bulletin of the History of Medicine* 79 (2005): 664–94.

Boddice, Rob. "The Manly Mind? Re-visiting the Victorian 'Sex in Brain' Debate." *Gender and History* 23 (2011): 321–40.

———. "Vivisecting Major: A Victorian Gentleman Scientist Defends Animal Experimentation, 1876–1885." *Isis* 102 (2011): 215–37.

———. "Species of Compassion: Aesthetics, Anaesthetics and Pain in the Physiological Laboratory." *19: Interdisciplinary Studies in the Long Nineteenth Century* 15 (2012).

———. "German Methods, English Morals: Physiological Networks and the Question of Callousness, c. 1870–1881." In *Anglo-German Scholarly Relations in the Long Nineteenth Century*, edited by Heather Ellis and Ulrike Kirchberger. Leiden, Netherlands and Boston: Brill, 2014.

———. "The Affective Turn: Historicising the Emotions." In *Psychology and History: Interdisciplinary Explorations*, edited by Cristian Tileagă and Jovan Byford. Cambridge, U.K.: Cambridge University Press, 2014.

———. *A History of Attitudes and Behaviours toward Animals in Eighteenth- and Nineteenth-Century Britain: Anthropocentrism and the Emergence of Animals*. Lewiston, N.Y.: Mellen, 2009.

Bourke, Joanna. *What it Means to be Human: Reflections from 1791 to the Present*. London: Virago, 2011.

———. *The Story of Pain: From Prayer to Painkillers*. Oxford, U.K.: Oxford University Press, 2014.

———. *Fear: A Cultural History*. London: Virago, 2005.

———. "Fear and Anxiety: Writing about Emotion in Modern History." *History Workshop Journal* 55 (2003): 111–33.

Bowler, Peter J. *Biology and Social Thought: 1850–1914*. Berkeley: Office for History of Science and Technology, University of California, 1993.

Brookes, Martin. *Extreme Measures: the Dark Visions and Bright Ideas of Francis Galton*. London: Bloomsbury, 2004.

Brunton, Deborah. *The Politics of Vaccination: Practice and Policy in England, Wales, Ireland, and Scotland, 1800–1874*. Rochester, N.Y.: University of Rochester Press, 2008.

Burdett, C. "Is Empathy the End of Sentimentality?" *Journal of Victorian Culture* 16 (2011): 259–74.

Chandler, James. *An Archaeology of Sympathy: The Sentimental Mode in Literature and Cinema*. Chicago: University of Chicago Press, 2013.

Clifford, David, ed. *Repositioning Victorian Sciences: Shifting Centres in Nineteenth-Century Scientific Thinking*. London: Anthem Press, 2006.

Csengei, Ildiko. *Sympathy, Sensibility and the Literature of Feeling in the Eighteenth Century*. Houndmills, U.K.: Palgrave, 2012.

Daston, Lorraine. "The Moral Economy of Science." *Osiris*, 2, no. 10 (1995): 2–24.

Daston, Lorraine, and Peter Galison. *Objectivity*. New York: Zone Books, 2007.

Dawson, Gowan. *Darwin, Literature and Victorian Respectability*. Cambridge, U.K.: Cambridge University Press, 2007.

Dixon, Thomas. *From Passions to Emotions: The Creation of a Secular Psychological Category*. Cambridge, U.K.: Cambridge University Press, 2003.

———. *The Invention of Altruism: Making Moral Meanings in Victorian Britain*. Oxford, U.K.: Oxford University Press, 2008.

———. *Weeping Britannia: Portrait of a Nation in Tears*. Oxford, U.K.: Oxford University Press, 2015.

Dror, Otniel. "The Affect of Experiment: The Turn to Emotions in Anglo-American Physiology, 1900–1940." *Isis* 90 (1999): 205–37.

———. "The Scientific Image of Emotion: Experience and Technologies of Inscription." *Configurations* 7 (1999): 355–401.

Durbach, Nadja. *Bodily Matters: The Anti-Vaccination Movement in England, 1853–1907*. Durham, N.C.: Duke University Press, 2005.

Elias, Norbert. *The Civilizing Process: Sociogenetic and Psychogenetic Investigations*, trans. Edmund Jephcott. 1939. New ed., Oxford, U.K.: Blackwell, 2000.

Endersby, Jim. *Imperial Nature: Joseph Hooker and the Practices of Victorian Science*. Chicago and London: University of Chicago Press, 2008.

Fichman, Martin. "Alfred Russel Wallace and Anti-vaccinationism in the Late Victorian Cultural Context, 1870–1907." In *Natural Selection and Beyond: The Intellectual Legacy of Alfred Russel Wallace*, edited by Charles H. Smith and George Beccaloni. Oxford, U.K.: Oxford University Press, 2008.

Fichman, Martin, and Jennifer E. Keelan. "Resister's Logic: The Anti-Vaccination Arguments of Alfred Russel Wallace and Their Role in the Debates Over Compulsory Vaccination in England, 1870–1907." *Studies in History and Philosophy of Biological and Biomedical Sciences* 38 (2007): 585–607.

Frazer, Michael. *The Enlightenment of Sympathy: Justice and the Moral Sentiments in the Eighteenth Century and Today*. Oxford, U.K.: Oxford University Press, 2012.

French, Richard D. *Antivivisection and Medical Science in Victorian Society*. Princeton, N.J.: Princeton University Press, 1975.

Frevert, Ute. *Emotions in History: Lost and Found.* Budapest: Central European University Press, 2011.

Gammerl, Benno, ed. *Emotional Styles—Concepts and Challenges.* Special issue of *Rethinking History* 16 (2012).

Gross, Daniel M. *The Secret History of Emotion: From Aristotle's Rhetoric to Modern Brain Science.* Chicago: University of Chicago Press, 2006.

———. "Defending the Humanities with Charles Darwin's *The Expression of the Emotions in Man and Animals* (1872)." *Critical Inquiry* 37 (2010): 34–59.

Himmelfarb, Gertrude. *Poverty and Compassion: The Moral Imagination of the Late Victorians.* New York: Vintage, 1991.

———. *The De-moralization of Society: From Victorian Virtues to Modern Values.* New York: Vintage, 1996.

Hochschild, Arlie Russell. "Emotion Work, Feeling Rules, and Social Structure." *American Journal of Sociology* 85 (1979): 551–75.

Jasanoff, Sheila, ed. *States of Knowledge: the Co-Production of Science and the Social Order.* London and New York: Routledge, 2004.

Kevles, Daniel J. *In the Name of Eugenics: Genetics and the Uses of Human Heredity.* Berkeley and Los Angeles: University of California Press, 1985.

Konstan, David. *Pity Transformed.* London: Duckworth, 2001.

Lansbury, Coral. *The Old Brown Dog: Women, Workers and Vivisection in Edwardian England.* Madison: University of Wisconsin Press, 1985.

Lightman, Bernard. *Evolutionary Naturalism in Victorian Britain: The "Darwinians" and their Critics.* Farnham, U.K.: Ashgate Variorum, 2009.

Lorimer, Douglas A. *Science, Race Relations and Resistance: Britain, 1870–1914.* Manchester: Manchester University Press, 2013.

Mazumdar, Pauline M.H. *Eugenics, Human Genetics and Human Failings: The Eugenics Society, its Sources and its Critics in Britain.* London: Routledge, 1992.

Mercer, Philip. *Sympathy and Ethics: A Study of the Relationship between Sympathy and Morality with Special Reference to Hume's Treatise.* Oxford, U.K.: Clarendon Press, 1972.

Miller, Ian. "Necessary Torture?: Vivisection, Suffragette Force-Feeding, and Responses to Scientific Medicine in Britain c. 1870–1920." *Journal of the History of Medicine and Allied Sciences* 64 (2009): 333–72.

Moore, James. "R. A. Fisher: A Faith Fit for Eugenics." *Studies in History and Philosophy of Biological and Biomedical Sciences* 38 (2007): 110–35.

Moscoso, Javier. *Pain: A Cultural History.* Houndmills, U.K.: Palgrave, 2012.

Nussbaum, Martha. *Upheavals of Thought: The Intelligence of Emotions.* Cambridge, U.K.: Cambridge University Press, 2003.

Olsen, Stephanie. *Juvenile Nation: Youth, Emotions and the Making of the Modern British Citizen, 1880–1914.* London: Bloomsbury, 2014.

Paul, Diane B., and James Moore. "The Darwinian Context: Evolution and Inheritance." In *The Oxford Handbook of the History of Eugenics,* edited by Alison Bashford and Philippa Levine, 27–42. Oxford, U.K.: Oxford University Press, 2010.

Pick, Daniel. *Faces of Degeneration: A European Disorder, c.1848-c.1918.* Cambridge, U.K.: Cambridge University Press, 1989.

Plamper, Jan. *The History of Emotions: An Introduction.* Oxford, U.K.: Oxford University Press, 2015.

———. "The History of Emotions: An Interview with William Reddy, Barbara Rosenwein, and Peter Stearns." *History and Theory* 49 (2010): 237–65.

Preece, Rod. "Darwin, Christianity and the Great Vivisection Debate." *Journal of the History of Ideas* 64 (2003): 399–419.

Prinz, Jesse. *The Emotional Construction of Morals.* Oxford, U.K.: Oxford University Press, 2007.

Porter, Dorothy, and Roy Porter. "The Politics of Anti-vaccinationism and Public Health in Nineteenth-century England." *Medical History* 32 (1988): 231–52.

Reddy, William. "Against Constructionism: The Historical Ethnography of Emotions." *Current Anthropology* 38 (1997): 327–51.

———. *The Navigation of Feeling: A Framework for the History of Emotions.* Cambridge, U.K.: Cambridge University Press, 2001.

Richards, Robert J. *Darwin and the Emergence of Evolutionary Theories of Mind and Behavior.* Chicago: University of Chicago Press, 1987.

Richards, Stewart. "Anaesthetics, Ethics and Aesthetics: Vivisection in the Late Nineteenth-Century British laboratory." In *The Laboratory Revolution in Medicine,* edited by Andrew Cunningham and Perry Williams, 142–69. Cambridge, U.K.: Cambridge University Press, 1992.

———. "Drawing the Life-Blood of Physiology: Vivisection and the Physiologists Dilemma, 1870–1900." *Annals of Science* 43 (1996): 27–56.

Richardson, Angelique. *Love and Eugenics in the Late Nineteenth Century: Rational Reproduction and the New Woman.* Oxford, U.K.: Oxford University Press, 2003.

Roberts, M.J.D. *Making English Morals: Voluntary Association and Moral Reform in England, 1787–1886.* Cambridge, U.K.: Cambridge University Press, 2004.

Rosenwein, Barbara. "Worrying about Emotions in History." *American Historical Review* 107 (2002): 821–45.

Rowbotham, Judith. "Legislating for Your Own Good: Criminalising Moral Choice. The Modern Echoes of the Victorian Vaccination Acts." *Liverpool Law Review* 30 (2009): 13–33.

Rowold, Katharina, ed. *Gender and Science: Late Nineteenth-Century Debates on the Female Mind and Body.* Bristol, U.K.: Thoemmes Press, 1996.

Rupke, Nicolaas, ed. *Vivisection in Historical Perspective.* London: Croom Helm, 1987.

Scheer, Monique. "Are Emotions a Kind of Practice (and Is That What Makes Them Have a History)? A Bourdieuian Approach to Understanding." *History and Theory* 51 (2012): 193–220.

Searle, G. R. *The Quest for National Efficiency: A Study in British Politics and Political Thought, 1899–1914.* Berkeley and Los Angeles: University of California Press, 1971.

Slotten, Ross A. *The Heretic in Darwin's Court: The Life of Alfred Russel Wallace.* New York: Columbia University Press, 2006.

Snow, Stephanie J. *Blessed Days of Anaesthesia: How Anaesthetics Changed the World.* Oxford, U.K.: Oxford University Press, 2008.

Stiles, Anne. *Popular Fiction and Brain Science in the Late Nineteenth Century.* Cambridge, U.K.: Cambridge University Press, 2012.

Stearns, Peter N., and Carol Z. Stearns. "Emotionology: Clarifying the History of Emotions and Emotional Standards." *American Historical Review* 90 (1985): 813–36.

Stearns, Peter, and Susan Matt, eds. *Doing Emotions History.* Urbana and Champaign: University of Illinois Press, 2014.

Szreter, Simon. "The Importance of Social Intervention in Britain's Mortality Decline c. 1850–1914: A Re-interpretation of the Role of Public Health." *Social History of Medicine* 1 (1988): 1–38.

Thomson, Mathew. *The Problem of Mental Deficiency: Eugenics, Democracy, and Social Policy in Britain c.1870–1959.* Oxford, U.K.: Clarendon Press, 1998.

Turda, Marius. *Modernism and Eugenics.* Houndmills, U.K.: Palgrave Macmillan, 2010.

Turner, Frank Miller. *Between Science and Religion: The Reaction to Scientific Naturalism in Late Victorian England.* New Haven, Conn., and London: Yale University Press, 1974.

Weisz, George. *Divide and Conquer: A Comparative History of Medical Specialization.* Oxford, U.K.: Oxford University Press, 2006.

White, Paul S. "Darwin's Emotions: The Scientific Self and the Sentiment of Objectivity." *Isis* 100 (2009): 811–26.

———. "The Experimental Animal in Victorian Britain." In *Thinking with Animals: New Perspectives on Anthropomorphism,* ed. Lorraine Daston and Gregg Mitman. New York: Columbia University Press, 2005.

———. Sympathy under the Knife: Experimentation and Emotion in Late Victorian Medicine." In *Medicine, Emotion and Disease, 1700–1950,* ed. Fay Bound Alberti. Houndmills, U.K.: Palgrave, 2006.

———, ed. "Focus: The Emotional Economy of Science." *Isis* 100 (2009): 792–851.

Williamson, Stanley. *The Vaccination Controversy: The Rise, Reign and Fall of Compulsory Vaccination for Smallpox.* Liverpool: Liverpool University Press, 2007.

Willis, Martin. "Unmasking Immorality: Popular Opposition to Laboratory Science in Late Victorian Britain." In *Repositioning Victorian Sciences: Shifting Centres in Nineteenth-Century Scientific Thinking,* edited by David Clifford. London: Anthem Press, 2006.

Winter, Alison. *Mesmerized: Powers of Mind in Victorian Britain.* Chicago: University of Chicago Press, 1998.

Index

ROB BODDICE works at the Department of History and Cultural Studies, Freie Universität Berlin. His books include *Edward Jenner* and *Pain: A Very Short Introduction*.

History of Emotions

Doing Emotions History Edited by Susan J. Matt and Peter N. Stearns
Driven by Fear: Epidemics and Isolation in San Francisco's House of Pestilence Guenter B. Risse
The Science of Sympathy: Morality, Evolution, and Victorian Civilization Rob Boddice

The University of Illinois Press
is a founding member of the
Association of American University Presses.

Cover designed by Jennifer Holzner
Cover image: Thomas Eakins, *Portrait of Dr. Samuel D. Gross (The Gross Clinic)*, oil on canvas, 1875.

University of Illinois Press
1325 South Oak Street
Champaign, IL 61820-6903
www.press.uillinois.edu

Printed and bound by CPI Group (UK) Ltd, Croydon, CR0 4YY

09/06/2025

14685835-0001